PROBLEMS BOOK and STUDY GUIDE

Nancy L. Pruitt
Colgate University

to accompany

CELL AND MOLECULAR BIOLOGY
Concepts and Experiments

Fourth Edition

Gerald Karp

Ch 10 – 8 –10 pm
Ch 11   10 – 12 pm
Ch 12   12 – 2 am
Ch 13   8 – 10
Ch 14   10 – 12
Ch 15   12 – 14
Ch 7   14 – 16
Ch 8   16 – 18
Ch 9   18 – 20
Ch 4   20 – 22
Ch 1   22 – 24
Ch 2   1 – 2
[Ch 18]

Ch 1, 2, 4, 9, 7, 15, 10, 11, 12, 13, 14, 8
2 korda ch

pühapäeval! kordus

WILEY

JOHN WILEY & SONS, INC.

To order books or for customer service, please call 1-800-CALL-WILEY (225-5945).

ISBN 0-471-46592-5

Printed in the United States of America

10 9 8 7 6 5 4 3 2 1

Printed and bound by Courier Kendallville, Inc.

To the Student

The astonishing pace at which new discoveries are being made in the fields of cell and molecular biology can be intimidating. You might ask yourself how the facts you learn in your cell biology course this year will look next year or a few years down the road. Will our understanding of cancer be the same? What new findings in cellular immunology will be made before I graduate from college? The best way to arm yourself against the obsolescence of facts is to construct your knowledge not around the details, but around concepts—the big ideas that form the scaffolding on which to hang the finer points. That way, when the facts grow or change, you will have the tools to integrate those changes into your understanding of the field. No doubt you will need to learn *some* facts, but to build lasting knowledge and become a life-long learner takes much more than memorization. You will need to develop your problem-solving skills using creating thinking; you will need to apply what you learn to novel situations; you will need to assess the value of various kinds of information, and the methods used to obtain it. That is what this workbook can help you to do. This kind of learning is more challenging than memorizing facts, but the extra effort is worth the trouble. The long term results are far more rewarding.

Here are some suggestions about how to get the most from and do well in your cell and molecular biology course: First, commit a certain number of hours each day to the study of cell and molecular biology. Do not let yourself get behind! Read each chapter of the text, then review the content using the outlines in this book called **Reviewing the Chapter**. You can glance only at the major headings in bold type, or allow your eye to take in more detail by reading subheadings. One of my favorite ways of testing my own understanding of a concept is to try explaining it to a friend. Practice explaining each of the **Learning Objectives**, anticipating questions. Can you answer them? If not, go back to the book or to the outline in **Reviewing the Chapter**.

Next, glance through the list of **Key Terms and Phrases**. They are arranged roughly in the order in which they appear in the chapters, so related terms will be close together. Do the terms conjure up mental images or ideas? If not, better check the text again, or **Reviewing the Chapter**.

Each chapter's **Key Figure** was selected to represent an important concept from the chapter. Use the **Questions for Thought** as a catalyst for discussion with a classmate, your professor, or teaching assistant, or just to ponder on your own.

Now get out your calculator and sharpen your pencils, you are ready to tackle the **Review Problems**. Try to answer the short answer questions in just a few carefully chosen words or phrases. If you consistently find that your answers take up more than the allotted space, perhaps you need to rethink your approach. Some of the multiple choice questions are simple recall, but others will ask you to take your knowledge one step further. Lastly, the **Problems and Essays** will require you to provide scientific explanations, interpret data, design experiments, or solve problems relating to the concepts in the chapter. Do your calculations right in the book, and don't cheat! Give yourself ample opportunity to work through to the answers on your own before looking in the **Appendix**.

Remember: The best professor, the best textbook, even the best study guide in the world cannot teach you unless you are committed to learning. Be an active participant in your own education.

Acknowledgments

I am privileged to teach cell and molecular biology to some of the finest students in the country, at Colgate University. Many of them have gone on to assume leadership roles in the field, and I hope I have played some small part in their success. I know my students play an important part in keeping me thinking about cell biology in new and productive ways, and I am grateful to them all—past and present.

I am also grateful to Project Editor Geraldine Osnato at John Wiley and Sons, who cheerfully answered my questions and ensured that I had the resources I needed.

A special thanks to my family, Jonathan, Nathan, and Danny, who made time for me when I needed time to work, made room for me to spread the chapters out all over the house, and made everything worthwhile for me in the end.

TABLE OF CONTENTS

Introduction to the Study of Cell and Molecular Biology

Learning Objectives

When you have finished this chapter, you should be able to:

1. Give a brief history of ideas about the cellular nature of life.
2. List the basic properties of cells.
3. Distinguish between prokaryotic and eukaryotic cells, and describe the characteristics of each, including:
   a. the major domains of prokaryotic cells.
   b. the process of differentiation and specialization in eukaryotes, including examples.
4. Appreciate the relative sizes of cells and their components, and the factors that limit the size of cells.
5. Understand the nature of viruses, their origins, and their mechanisms of infection.
6. Describe some other noncellular pathogens, including virions and prions.
7. Become familiar with the theory and evidence that eukaryotic cells arose from an endosymbiotic relationship between prokaryotic ancestors.

Key Terms and Phrases

Robert Hooke
HeLa cells
metabolism
eukaryote
nucleus
mitochondria
cytoskeleton
Eucarya
methanogens
cyanobacteria
differentiation
virus
provirus
endosymbiont theory
lateral gene transfer

Anton van Leeuwenhoek
Schleiden, Schwann and Virchow
organelles
cell wall
nuclear envelope
endoplasmic reticulum
ribosome
Eubacteria
halophile
nitrogen fixation
micrometer
virion
viroid
Lynn Margulis

cell theory
ATP
prokaryote
cell surface receptors
nucleoid
chloroplast
bacterial conjugation
Archaea
thermoacidophile
Dictyostelium
nanometer
capsid
prion
Woese and Fox

Reviewing the Chapter

I. Introduction
  A. Cells are the topic of intense study.
  B. The study of cells requires creative instruments and techniques.
  C. The study of cells lies on a spectrum of human endeavor, from subatomic physics to astronomy.

II. The Discovery of Cells
  A. The discovery of cells followed from the invention of the microscope by Robert Hooke, and its refinement by Anton van Leeuwenhoek.
  B. Cell theory was articulated in the mid-1800s by Schleiden, Schwann and Virchow.
    1. All organisms are composed of one or more cells.
    2. The cell is the structural unit of life.
    3. Cells arise from preexisting cells by division.

III. Basic Properties of Cells
  A. Life is the most basic property of cells.
  B. Cells can grow and reproduce in culture for extended periods.
    1. HeLa cells are cultured tumor cells isolated from a cancer patient named Henrietta Lacks by George Gey in 1951.
    2. Cultured cells, including HeLa cells, are an essential tool of cell biologists.
  C. Cells are highly complex and organized.
    1. Cellular processes are highly regulated.
    2. Cells from different species share similar structure, composition and metabolic features that have been conserved throughout evolution.
  D. Cells have a genetic program.
    1. Genes encode information to build each cell, and the organism.
    2. Genes encode information for cellular reproduction, activity, and structure.
  E. Cells reproduce, and each daughter cell receives a complete set of genetic instructions.
  F. Cells acquire and utilize energy.
    1. Photosynthesis provides fuel for all living organisms.
    2. Animal cells derive energy from the products of photosynthesis, often in the form of glucose.
    3. Cells can convert glucose into ATP— a substance with readily available energy.
    4. Cells perform chemical reactions catalyzed by enzymes.
      (a) Chemical reactions occur in sequences.
      (b) Enzymes increase the rate of cellular chemical reactions.
    5. Metabolism is the sum of all cellular chemical reactions.
  G. Cells engage in numerous mechanical activities, including locomotion.
  H. Cells are able to respond to stimuli using surface receptors that sense changes in the chemical environment.
  I. Cells are capable of self-regulation.

IV. Two Fundamentally Different Classes of Cells
  A. Prokaryotes are all bacteria. They arose more than 3.7 billion years ago.
  B. Eukaryotes are protists, fungi, plants and animals. The first eukaryotes arose 1.6 billion years ago.
  C. Characteristics that distinguish prokaryotic and eukaryotic cells:
    1. Complexity: Prokaryotes are relatively simple; eukaryotes are more complex in structure and function.
    2. Genetic material:

        a. Packaging: Prokaryotes have a nucleoid region whereas eukaryotes have a true membrane-bound nucleus.
        b. Amount: Eukaryotes have several orders of magnitude more genetic material than prokaryotes.
        c. Form: Eukaryotes have many chromosomes that are made of both DNA and protein whereas prokaryotes have a single, circular DNA chromosome.
    3. Cytoplasm: Eukaryotes have membrane-bound organelles and complex cytoskeletal proteins; prokaryotic cytoplasm is devoid of membranous structures and contains primitive cytoskeletal filaments. Both have ribosomes, although they differ in size.
    4. Cellular reproduction: Eukaryotes divide by mitosis; prokaryotes divide by simple fission.
    5. Locomotion: Eukaryotes use both cytoplasmic movement, and cilia and flagella; prokaryotes have flagella, but they differ in both form and mechanism from eukaryotic flagella.

V. Types of Prokaryotic Cells: Two Domains
  A. Domain Archaea:
    1. Methanogens
    2. Halophiles
    3. Acidophiles
    4. Thermophiles
  B. Domain Bacteria:
    1. Includes the smallest known cells— the mycoplasma.
    2. Includes cyanobacteria— some photosynthetic bacteria
        a. Cyanobacteria gave rise to green plants and an oxygen-rich atmosphere.
        b. Some cyanobacteria are nitrogen fixers.
  C. Prokaryotic Diversity
    1. Prokaryotes are identified and classified on the basis of specific DNA sequences.
    2. Recent evidence indicates that prokaryotes are more diverse and numerous than previously imagined.

VI. Types of Eukaryotic Cells: Cell Specialization
  A. Unicellular eukaryotes are complex single-celled organisms.
  B. Multicellular eukaryotes have different cell types for different functions.
    1. The slime mold *Dictyostelium* is an example of unspecialized cells becoming specialized via differentiation during the life cycle.
    2. Differentiation occurs during embryonic development in other multicellular organisms.
    3. Numbers and arrangements of organelles relate to the function of the cell.
    4. Despite cellular differentiation, eukaryotic cells have many features in common.
    5. Model Organisms:
        a. Cell research focuses on five model organisms.
        b. These are the yeast *Saccharomyces*, the mustard plant *Arabidopsis*, the nematode *Caenorhabditis elegans*, the fruit fly *Drosophila* and the mouse *Mus*.

VII. The Human Perspective: The Prospect for Cell Replacement Therapy
  A. Stem cells are undifferentiated cells capable of self-renewal and differentiation.
    1. Adult stem cells can be used to replace damaged or diseased adult tissue.
        a. Hematopoietic stem cells can produce blood cells in bone marrow.
        b. Neural stem cells may be used to treat neurodegenerative diseases.

2. Embryonic stem (ES) cells have even greater potential for differentiation than adult stem cells.
   a. ES cells must be differentiated in vitro, lest they cause cancer.
   b. The use of ES cells involves ethical considerations.
B. Organ replacement in the future may involve growing organs from stem cells.
   1. Organs must be grown on inert scaffolding in the laboratory.
   2. Stem cells used for organ regeneration must be compatible with the recipient's immune system.
   3. Growing organs for transplant is still experimental.

VIII. The Size of Cells and Their Components

A. Cells are commonly measured in units of micrometers (1 μm = 10^{-6} meter) and nanometers (1 nm = 10^{-9} meter).
B. Cells size is limited:
   1. by the volume of cytoplasm that can be supported by the genes in the nucleus.
   2. by the volume of cytoplasm that can be supported by exchange of nutrients across a limited surface area, i.e., the surface area/volume ratio.
   3. by the distance over which substances can efficiently travel through the cytoplasm via diffusion.

IX. Viruses

A. Viruses are pathogens first described in the late 1800s.
B. Viruses are obligatory intracellular parasites.
C. A virion is a virus particle outside the host cell.
D. Viral structure:
   1. The genetic material can be single- or double-stranded DNA or RNA.
   2. The protein capsid surrounds the genetic material in the virion.
   3. The capsid may be polyhedral in some viruses.
   4. A lipid envelope may surround the capsid in some viruses; the lipids may contain viral proteins.
E. Virus and host:
   1. Viruses have surface proteins that bind to the surface of the host cell.
   2. Viral specificity for a certain host is determined by the virus's surface proteins.
F. Viruses are not considered living.
G. Viral infection types:
   1. Lytic infection— the virus redirects the host into making more virus particles, the host cell ruptures and releases the viruses.
   2. Integrated infection— the virus integrates its DNA (called the provirus) into the host cell's chromosomes.
      a. The infected host may behave normally until an external stimulus activates the provirus, leading to lysis and release of viral progeny.
      b. The host may give rise to viral progeny by bidding, as in HIV.
      c. The host may become malignant.
H. Viral origins:
   1. Viruses had to arise after their hosts evolved— they cannot survive without their hosts.
   2. Viruses probably arose as fragments of host chromosomes that became somewhat autonomous.
I. Viruses are good models for studying mechanisms of genetic expression, and good vectors for introducing foreign genes into cells.
J. The journey of viruses through living cells has recently been studied using fluoescent molecules.

X. Viroids
    A. Viroids are pathogens, each consisting of a small, naked RNA molecule.
    B. Viroids cause disease by interfering with gene expression in host cells.

XI. Experimental Pathways: The Origin of Eukaryotic Cells
    A. Prokaryotic cells arose first and gave rise to eukaryotic cells.
    B. Endosymbiont Theory: organelles in eukaryotic cells (mitochondria and chloroplasts) evolved from smaller prokaryotic cells.
    C. Evidence to support the Endosymbiont theory
        1. Absence of eukaryote species with organelles in an intermediate stage of evolution.
        2. Many symbiotic relations are known among different organisms.
        3. Organelles of eukaryotic cells contain their own DNA and the machinery to synthesize proteins.
        4. The nucleotide sequences of rRNAs from eukaryotic organelles resembled that of prokaryotes.
        5. Organelles duplicate independently of the cell nucleus.
    D. New ideas about the origin of eukaryotic cells have emerged from studies of genomes.
        1. Based on nucleotide sequences of single genes, Carl Woese proposed three major cell lineages.
            a. Bacteria include gram positive, gram negative and cyanobacteria.
            b. Archaea include halophiles, thermophiles, methangens and acidophiles.
            c. Eucarya include plants, animals, fungi and protists.
        2. Based on whole genomes, many organisms appear to be genetic mosaics.
            a. Lateral gene transfer results in organisms with both parental DNA and DNA from other organisms in the environment.
            b. Bacteria and Eucarya show evidence of lateral gene transfer in their genomes.

Key Figure

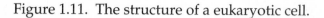

Figure 1.11. The structure of a eukaryotic cell.

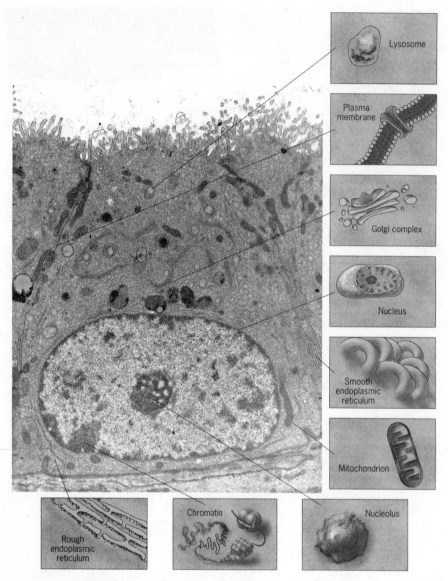

Questions for Thought:

1. Is this a plant cell or an animal cell? What structures identify it as such?
2. What structures shown here do plant cells, animal cells and prokaryotic cells have in common?
3. What is the largest intracellular structure in the plant cell? The animal cell? The bacterial cell?
4. This is an epithelial cell, shown to give you an overview of what kinds of structures occur in this cell type. How might the numbers of the different organelles differ in cells with specific functions?

Review Problems

<u>Short Answer</u>

1. Give the significance (not the definition) of the following terms or phrases. Say what they do or why they are important. For example:
photosynthesis: *ultimate source of energy supporting all life on Earth.*

    a. cells

    b. ATP

    c. model organisms

    d. prokaryotes and eukaryotes share the same genetic "language"

    e. nitrogen fixation

    f. surface proteins on viruses

    g. genetic similarity between genes of viruses and their hosts

    h. life cycle of *Dictyostelium*

    i. bacterial conjugation

2. Compare and contrast the following:

    a. prokaryote vs. eukaryote

    b. virus vs. virion

    c. lytic infection vs. integrative infection

    d. Bacteria vs. Archaea

Multiple Choice

1. All of the following individuals contributed to cell theory except:
    a. Robert Hooke.
    b. Matthias Schleiden.
    c. Theodor Schwann.
    d. Rudolf Virchow.

2. Of the following structures, which is the smallest?
    a. viroid
    b. hydrogen atom
    c. bacterium
    d. mitochondrion

3. Of the following, which is the most primitive?
    a. virus
    b. eukaryote
    c. Archaeon
    d. mitochondria

4. Cell theory includes all of the following except:
    a. All organisms are composed of one or more cells.
    b. The cell is the most primitive form of life.
    c. The cell is the structural unit of life.
    d. Cells arise by division of preexisting cells.

5. All of the following are basic properties of cells except:
    a. Cells have nuclei and mitochondria.
    b. Cells have a genetic program and the means to use it.
    c. Cells are capable of producing more of themselves.
    d. Cells are able to respond to stimuli.

6. The Archaea include all of the following except:
   a. methanogens.
   b. halophiles.
   c. thermoacidophiles.
   d. cyanobacteria.

7. Evolutionary relationships between groups of organisms are determined using which of the following types of information?
   a. comparisons of nucleotide sequences
   b. comparisons of biochemical pathways
   c. comparisons of structural features
   d. all of the above

8. All of the following are features of prokaryotes except:
   a. nitrogen fixation.
   b. photosynthesis.
   c. sexual reproduction.
   d. locomotion.

9. Which of the following may account for the small size of cells?
   a. the rate of diffusion
   b. the surface area/volume ratio
   c. the number of mRNAs that can be produced by the nucleus
   d. all of the above

10. Which of the following statements is not true of viruses?
    a. Viruses have been successfully grown in pure cultures in test tubes.
    b. All viruses are obligatory intracellular parasites.
    c. All viruses have either DNA or RNA as their genetic material.
    d. Viruses probably arose from small fragments of cellular chromosomes.

11. If you were to study the sequences of nucleic acids in a variety of viruses and viral hosts, you would probably find more similarities:
    a. among different viruses than between viruses and their hosts.
    b. among different viral hosts than among different viruses.
    c. among different viral hosts than between viruses and their hosts.
    d. between viruses and their hosts than among different viruses.

12. Bone marrow transplants are to blood transfusions as organ engineering is to __?__ .
    a. organ transplant.
    b. organ transfusion.
    c. bone marrow transfusions.
    d. stem cell growth.

13. The *most* powerful tool for determining evolutionary relationships among cell types is
    a. the study of cell shape.
    b. the study of organelle shape.
    c. the study of amino acid sequences in proteins.
    d. the study of nucleic acid sequences in genomes.

Problems and Essays

1. Mitochondria and chloroplasts can be isolated from broken cells and, given the proper nutrients, can undergo respiration and photosynthesis in a test tube for several hours. Why, then, are these organelles not considered the basic unit of life?

2. Given that all the cells of the slime mold *Dictyostelium* have identical genetic information, why do some develop into stalk cells and others into spore cells?

3. Given that cell structure reflects cell function, what structural features would you predict in the following?

    a. cells that line the digestive tract and take up nutrients across the wall of the intestine into the blood

    b. cells that synthesize and secrete protein, such as an insulin-producing cell from the pancreas or an antibody-producing lymphocyte

    c. a leaf cell

    d. a contractile cell, such as a muscle cell

4.   Of viruses that infect bacteria or viruses that infect humans, which type is older and why?

5.   Consider a large, roughly cuboidal cell that measures 100 µm on a side.
     a.   What is its surface area/volume ratio?

     b.   Assume that a cell requires a surface area/volume ratio of at least 3 to survive. Would dividing this cell into 125 cells which together had a volume of 1,000,000 µm3 ensure survival?

6. Given that some antibiotic drugs kill pathogens by interfering with replication of bacterial chromosomes, why can't we use these drugs to cure the common cold?

7. Why does the HIV virus lack genes for synthesizing its own lipid envelope?

8. What biochemical evidence could be used to show that modern green plants evolved from cyanobacterial ancestors and not from other photosynthetic prokaryotes?

9. Imagine that you have just discovered a new disease that infects the roots of corn plants and threatens the corn crop in the United States. You have been given a major grant from the U. S. Department of Agriculture to study this plant disease. Begin your studies by answering the following two questions:

   a. What kinds of pathogens might be causing this disease?

   b. What experiments will you do to distinguish among these different pathogens?

Web Explorations

Mycoplasmas are the smallest of the prokaryotic cells—indeed the smallest of all living cells. Their role in some human health problems is still controversial. Some claim that mycoplasmas are responsible for several enigmatic diseases, including Crohn's disease, arthritis, Gulf War syndrome, chronic fatigue syndrome, and even some forms of cancer.

Using the World Wide Web, choose one of the diseases mentioned above and look for a possible role for mycoplasmas in that disease. Remember to be critical of what you read on the web. The information on the web is not always evaluated for accuracy; many web sites are posted by groups with specific agendas or decidedly unscientific ideas. Check to see who has posted each site you use, and make an informed decision about whether that information can be trusted. Decide for yourself if there is enough scientific evidence to link mycoplasmas to that disease.

2 The Chemical Basis of Life

Learning Objectives

When you have finished this chapter, you should be able to:

1. Explain why an understanding of chemistry is essential to the study of cell biology.
2. Describe the nature of covalent bonds, and why some bonds have unequal distributions of electrical charge.
3. Identify polar and nonpolar molecules, and understand the consequences of polarity in cellular molecules.
4. List the types of weak bonds and explain why they are important in cells.
5. Explain why water is ideally suited to life.
6. Define and understand the importance of pH in cells.
7. List the four cellular macromolecules and their building blocks. Describe how the building blocks are assembled into macromolecules.
8. Describe the levels of protein structure, and the factors that determine the structure of proteins in cells.
9. Give examples of some important cellular proteins.

Key Terms and Phrases

molecule	compound	covalent bond
electronegativity	polar	nonpolar
ionization	anion	cation
free radical	weak bond	ionic bond
hydrogen bond (H-bond)	hydrophilic	hydrophobic
van der Waals forces	metabolic intermediate	carbohydrate
carbonyl group	ketose	aldose
stereoisomer (enantiomer)	asymmetric carbon	pyranose
glycosidic bond	disaccharide	oligosaccharide
glycogen	starch	amylose
amylopectin	dissociation	pH
amphoteric	acid	base
K_w	buffer	organic molecule
inorganic molecule	biochemical	hydrocarbon
saturated hydrocarbon	unsaturated hydrocarbon	structural isomer
functional group	ester bond	amide bond
macromolecule	protein	amino acid
polysaccharide	oligosaccharide	sugar
polymer	monomer	nucleic acid
nucleotide	cellulose	chitin
glycosaminoglycan	cellulase	lipid
fatty acid	oil	triacylglycerol
amphipathic	adipocyte	steroid
atherosclerosis	phospholipid	diacylglycerol
polypeptide chain	peptide bond	C-terminus

N-terminus	R group (side chain)	disulfide bridge
conjugated protein	nucleoprotein	lipoprotein
glycoprotein	primary structure	secondary structure
tertiary structure	quarternary structure	sickle cell anemia
conformation	alpha helix (α helix)	beta sheet (ß sheet)
fibrous protein	globular protein	keratin
collagen	x-ray crystallography	myoglobin
prosthetic group	protein domain	protein motif
homodimer	heterodimer	subunit
denaturation	isoforms	homologous
site-directed mutagenesis	deoxyribonucleic acid (DNA)	ribonucleic acid (RNA)
purine	pyrimidine	adenine
guanine	cytosine	uracil
thymine	adenosine triphosphate (ATP)	self-assembly
heat shock response	molecular chaperone	

Reviewing the Chapter

I. Introduction
    A. Understanding cellular function requires knowledge of structure.
    B. Structure and function in cells is closely related to the structure of molecules and atoms.
    C. The study of chemistry is essential for understanding cell biology.

II. Covalent Bonds
    A. Bonds between atoms with shared pairs of electrons are called covalent bonds.
        1. Molecules are stable combinations of atoms held together by covalent bonds.
        2. Compounds are molecules with more than one type of atom.
    B. Atoms tend to fill their outer electron shell by sharing electrons with other atoms.
        1. Hydrogen forms a single covalent bond by sharing its one unpaired electron.
        2. Oxygen forms two covalent bonds by sharing two unpaired electrons.
        3. Water results when oxygen bonds with two hydrogens.
    C. It requires 80 to 100 kilocalories to break a mole of covalent bonds.
    D. Multiple bonds are formed when more than one pair of electrons are shared by two atoms.
    E. Shared electrons stay closest to the nucleus with the highest electronegativity.
        1. The electronegativity depends upon the number of protons in the nucleus.
        2. The electronegativity also depends upon the distance of the shared electrons from the nucleus.

III. Polar and Nonpolar Molecules
    A. Polar molecules have asymmetric distributions of electrical charge.
    B. Nonpolar molecules lack polarized bonds.
    C. Some biological molecules, such as proteins and phospholipids, have both polar and nonpolar regions and are called "amphipathic" molecules.

IV. Ionization
    A. Ions result when strongly electronegative nuclei capture electrons.
    B. Anions have extra electrons; cations have lost electrons.
    C. Free radicals are unstable atoms or molecules with unpaired electrons.

V. The Human Perspective: Free Radicals as a Cause of Aging
    A. Free radicals are atoms or molecules with an unpaired electron.

1. Free radicals are formed during normal metabolism.
2. Free radicals are highly reactive and damage macromolecules such as DNA..
3. Free radicals may play a role in aging.
   B. Superoxide dismutase (SOD) is an enzyme that destroys the superoxide radical $(O_2 \bullet)$.
      1. SOD protects cells from damage due to the superoxide radical.
      2. SOD extends the life span of fruit flies that overproduce it.
   C. Calorie restriction:
      1. Extends the lifespan of experimental mammals
      2. Results in decreased production of free radicals.
   D. Antioxidants destroy free radicals.
      1. Foods rich in natural antioxidants are highly beneficial.
      2. Antioxidant supplements have not increased the life span in studies with rodents.

VI. **Noncovalent Bonds**
   A. Noncovalent bonds are attractive forces that are weaker than covalent bonds (about 1 to 5 kcal/mole).
   B. Weak bonds are readily broken and re-formed.
   C. Weak bonds play an important role in dynamic cellular processes.
   D. Types of noncovalent bonds
      1. Ionic bonds: Attraction between charged atoms:
         a. Ionic bonds are weakened in the presence of water.
         b. They may be quite important within large biological molecules.
      2. Hydrogen bonds:
         a. Hydrogen bonds occur when covalently bound hydrogen has a partial positive charge and attracts electrons of a second atom.
         b. H-bonds determine the structure and properties of water.
         c. H-bonds occur in biological molecules, notably between the strands of DNA.
      3. Hydrophobic interactions and van der Waals forces:
         a. These occur when nonpolar molecules associate and minimize their exposure to polar molecules.
         b. Van der Waals forces, or attractions between nonpolar molecules, are due to transient dipole formation.

VII. **The Life Supporting Properties of Water**
   A. The structure of water is suitable for sustaining life.
      1. The molecule is asymmetric– both H atoms are on one side.
      2. Both covalent O–H bonds are highly polarized.
      3. All three atoms readily form hydrogen bonds.
   B. The properties of water result from its structure.
      1. Water requires a lot of heat energy to raise its temperature.
      2. Water requires a lot of heat to evaporate it.
      3. Water is an excellent solvent for many substances.
      4. Water determines the interactions between many biological solutes.

VIII. **Acids, Bases and Buffers**
   A. Acids release protons; bases accept protons. $H_3O^+ + OH^- \rightarrow H_2O + H_2O$
   B. Amphoteric molecules can act as either acids or bases; water is an example of an amphoteric molecule.
   C. Acidity is measured using the pH scale.
      1. $pH = -\log[H^+]$
      2. The ion product constant of water, $K_W = [H^+][OH^-] = 10^{-14}$ at 25 °C.

3. As [H⁺] increases, then [OH⁻] decreases such that the product of the two always equals 10^{-14}.
   D. Biological processes are sensitive to pH.
      1. Changes in pH affect the ion state and function of proteins.
      2. Buffers in living systems, such as bicarbonate, resist changes in pH.

IX. The Nature of Biological Molecules
   A. Carbon is central to the chemistry of life.
      1. Carbon forms four covalent bonds, with itself or other atoms.
      2. Several hundred thousand carbon-containing (organic) molecules have been identified.
   B. Hydrocarbons:
      1. Hydrocarbons contain only carbon and hydrogen.
      2. Hydrocarbons vary in the number of carbons, and the number of double and triple bonds between carbons.
      3. Structural isomers have the same chemical formula, but the atoms are arranged differently.
   C. Functional groups are groups of atoms that behave as units giving organic molecules different characteristics and properties.
   D. Functional classification of biological molecules:
      1. Macromolecules:
         a. Macromolecules are large structural and functional molecules in cells.
         b. Macromolecules include proteins, nucleic acids, polysaccharides and lipids.
      2. Building blocks of macromolecules include amino acids for proteins, nucleotides for nucleic acids, sugars for polysaccharides, and fatty acids for lipids.
      3. Metabolic intermediates are compounds formed in metabolic pathways.
      4. Miscellaneous molecules include vitamins, hormones, ATP, and metabolic waste products.

X. A Classification of Biological Molecules by Function
   A. Macromolecules are huge molecules that form the structure and carry out the functions of cells.
   B. The building blocks of macromolecules are low molecular weight precursors of macromolecules.
   C. Metabolic intermediates are compounds formed along metabolic pathways.
   D. Molecules of miscellaneous function include vitamins, hormones, ATP, regulatory molecules and wastes.

XI. Four Types of Macromolecules
   A. Carbohydrates include simple sugars and sugar polymers.
      1. Carbohydrates serve as energy storage molecules.
      2. Structure:
         a. Chemical formula is $(CH_2O)_n$.
         b. Ketose sugars have a carbonyl on an internal carbon.
         c. Aldose sugars have a carbonyl on a terminal carbon.
         d. Sugars can be linear, but usually form ring structures called pyranoses.
      3. Stereoisomerism:
         a. Asymmetric carbons bond to four different groups.
         b. Molecules with asymmetric carbons can exist in two mirror-image configurations called enantiomers or stereoisomers.
         c. Enantiomers are designated either D- or L-isomers.

(margin handwritten notes, top):
$-\overset{O}{C}-O-$ ester $C-O-C$ ether $C-OH$ alcohol $-\overset{O}{C}-OH$ acid $C-C-C$ ketone $\overset{O}{C}-CH$ aldeh. HCH hotemaldeh

d. Sugars can have many asymmetric carbons, but are designated D- or L- according to the arrangement around the carbon farthest from the carbonyl group.

4. Linking sugars together:
   a. Glycosidic bonds are C–O–C links between sugars.
   b. Disaccharides are used as a source of readily available energy.
   c. Oligosaccharides are found bound to cell surface proteins and lipids, and may be used for cell recognition.

5. Polysaccharide are polymers of sugars joined by glycosidic bonds.
   a. Glycogen and starch: nutritional polysaccharides.
     (1) Glycogen is an animal product made of branched glucose polymers.
     (2) Starch is a plant product made of both branched and unbranched glucose polymers.
   b. Cellulose, chitin, and glycosaminoglycans: structural polysaccharides.
     (1) Cellulose is a plant product made of unbranched glucose polymers that most animals cannot digest.
     (2) Chitin is a component of invertebrate exoskeleton made of unbranched N-acetylglucosamine polymers.
     (3) Glycosaminoglycans (GAGs) are composed of two different sugars and are found in the extracellular space.

B. Lipids are a diverse group of nonpolar molecules.
  1. Fats are made of glycerol molecules esterified to fatty acids.
   a. Fatty acids are unbranched hydrocarbons with one terminal carboxyl group; they are amphipathic.
   b. Saturated fatty acids lack C=C double bonds and are solids at room temperature.
   c. Unsaturated fatty acids have one or more C=C double bonds and are liquid at room temperature.
   d. In animals, fats are stored for energy in adipocytes.
  2. Steroids are four-ringed animal lipids that have been implicated in atherosclerosis.
  3. Phospholipids are amphipathic lipids that are a major component of cell membranes.

C. Proteins are polymers of amino acids and form a diverse group of macromolecules with a variety of cellular functions.
  1. Amino acids all have an alpha (α) carbon, an amine group, a carboxyl group, and a variable R group.
  2. Amino acids in nature occur as the L stereoisomer.
  3. Amino acids are linked together by peptide bonds into polypeptide chains to make proteins.
  4. Peptide bonds form between the α -carboxyl and the α -amino of participating amino acids.
  5. Amino acids differ in the R group attached to one of the bonds of the α - carbon.
  6. Properties of the side chains (R Groups):
   a. R groups may be polar and charged.
   b. R groups may be polar and uncharged.
   c. R groups may be nonpolar.
   d. R groups may have other properties.
     (1) Glycine has only -H as its R group and is small.
     (2) Proline's α -carbon is part of a ring, creating kinks in the protein.
     (3) Cysteine forms disulfide bonds with other cysteines.

(margin handwritten notes, lower left):
$-\overset{O}{C}-\underset{H}{N}-$ Peptide bond.

gly   pro   cys $-CH_2-S-S-CH_2-$

e. The nature of the R groups determines the function of the protein.
7. Conjugated proteins have nonprotein constituents.
8. The Structure of Proteins:
    a. Primary structure, the sequence of amino acids in the polymer, is critical to protein function.
    b. Secondary structure refers to the conformation of adjacent amino acids into α -helix, ß-sheet, hinges, turns, loops or finger-like extensions.
    c. Tertiary structure is the conformation of the entire polymer.

Polar charged:

Asp , Glu , Lys, Arg , His

        (1) Tertiary structure is stabilized by noncovalent bonds.
        (2) Tertiary structure is studied using x-ray crystallography.
        (3) Proteins can be fibrous or globular.
        (3) Myoglobin: The first globular protein whose tertiary structure was determined:

Polar uncharged:

Ser, Thr , Glu, Asn , Tyr

            (a) Myoglobin stores oxygen in muscle cells.
            (b) Myoglobin has a heme prosthetic group that binds oxygen.
            (c) Myoglobin structure was derived using x- ray crystallography, and found to be predominantly α -helix.

Nonpolar:

Ala , Val , Leu , Ile , Met , Phe , Trp

        (4) Protein domains and motifs:
            (a) Domains occur when large proteins are composed of more than one distinct functional region.
            (b) Motifs are recurring substructures in different proteins that may be related to similar evolutionary histories or similar functions.
        (5) Dynamic changes within a protein may occur with protein activity.
    d. Quarternary structure applies to proteins with subunits, and refers to the manner in which subunits interact.
9. Protein structures can be greatly infuenced by the energy of their environment.
10. Multiprotein complexes occur when groups of proteins associate to perform a specific function.
11. Protein-protein interaction can be studied by assays such as the Y2H system.
12. Protein folding is a process that occurs in orderly steps.
    a. Two theories of proper protein folding have been proposed:
        (1) Proteins may assume their native conformation via a series of orderly steps along a defined pathway, each of which enhances the next step.
        (2) Proteins may fold along smooth or bumpy "landscape funnels" in which there are no specific intermediate forms.

Sheat ley

        (3) Smaller proteins with single domains fold faster than larger proteins with multiple domains.
    b. Some proteins require molecular chaperones to prevent nonselective interactions while they fold.
        (1) Hsp70 bind nacent proteins and prevent inappropriate interactions.
        (2) Chaperonins sequester some larger new proteins in a central cavity to enable proper folding.
        (3) Hsp90 may enable mutant proteins to function normally; Hsp90 also protects against heat denaturation.
D. The Emerging Field of Proteomics:
    1. The proteome is the entire collection of an organism's proteins.

2. Proteomics uses advanced technologies to perform systematic, large scale studies on diverse mixtures of proteins.
    a. Proteins are separated using gel electrophoresis.
    b. Proteins are identified using mass spectrometry and high speed computers.
3. Protein microarrays enable scientists to screen proteins for various activities and disorders.
4. Proteonomics is widly used in pharamaceutical and medical industries to identify the role of protein disorders in diseases.
E. Protein engineering:
    1. Current technology enables us to make artificial genes that code for proteins of specified amino acid sequences.
    2. Knowledge of a protein's amino acid sequence rarely permits us to predict a protein's higher-order structure.
    3. Site-directed mutagenesis enables researchers to make alterations in single amino acids of a protein by altering the DNA that encodes it.
    4. Structure-Based Drug Design:
        a. Computer simulations of protein binding sites are used to design drugs that inhibit specific proteins.
        b. Non-aspirin anti-inflammatory drugs, such as Celebrex and Vioxx, target a specific cyclooxygenase, functioning like aspirin but without side effects.
F. Protein Adaptation and Evolution:
    1. Adaptations are traits that improve the chance of survival of an organism in a particular environment.
    2. Proteins are subject to natural selection.
    3. Members of a protein family are thought to have evolved from a single ancestral gene.
    4. Mammalian proteins tend to be larger and contain more domains than proteins of less complex organisms.
G. Nucleic acids are polymers of nucleotides that store and transmit genetic information.
    1. DNA holds the genetic information in all cellular organisms and some viruses.
    2. RNA holds the genetic information in other viruses.
    3. Nucleotides consist of a five-carbon sugar (ribose in RNA), a phosphate group and a nitrogenous base.
    4. Nucleotides are connected via 3'-5' phosphodiester bonds between the phosphate of one nucleotide and the 3' carbon of the next.
    5. The nitrogenous bases are either purines or pyrimidines.
        a. The purines are adenine and guanine in both DNA and RNA.
        b. The pyrimidines are cytosine and uracil in RNA and cytosine and thymine in DNA.
    6. RNA is usually single stranded and DNA is usually double stranded.
        a. RNA may fold back on itself to form complex three dimensional structures, as in ribosomes.
        b. RNA may have catalytic properties; such RNAs are called ribozymes.
    7. ATP is a ribonucleotide with three phosphates, and plays an important role in cellular metabolism.

XII. The Formation of Complex Molecular Structures
A. Different types of subunits are capable of self-assembly to form higher-order arrangements.
B. The assembly of tobacco mosaic virus particles and ribosomal subunits:

1. Experiments in the 1950s showed that TMV particles are capable of self-assembly.
2. Prokaryotic ribosomal subunits were shown in the 1960s and 1970s to assemble spontaneously in a step-by-step manner in vitro.
3. The components of the mature eukaryotic ribosomal subunits require accessory proteins to assemble.
   C. Cells may use chaperone proteins to assemble molecular structures.

XIII. The Human Perspective: Protein Misfolding Can Have Deadly Consequences

A. Creutzfeld-Jakob disease (CJD) results from a misfolded protein in the brain.
   1. Healthy brains contain a normal protein, PrP^c.
   2. CJD brains have a version of PrP^c, called PrP^{Sc}, that is identical or similar in amino acid sequence, but folded to contain primarily β sheets.
   3. The abnormal PrP^{Sc} protein, also called a prion, causes normal PrP^c to misfold.
   4. "Mad cow disease," kuru, and scrapie are also caused by PrP^{Sc}.
B. Alzheimer's disease involves misfolded proteins that accumulate in the brains of afflicted individuals.
   1. A membrane-bound protein in brain neurons, called amyloid precursor protein, is cleaved by two secretase enzymes.
   2. In individuals genetically predisposed to Alzheimer's disease one of the cleavage products is Aβ42, a protein that misfolds and self-associates into amyloid plaques.
   3. Several treatment therapies, including immunization with Aβ42, show promise.

XIV. Experimental Pathways: Chaperones: Helping Proteins Reach their Proper Folded State

A. Some proteins can self-assemble from purified subunits.
B. Other proteins require molecular chaperones for proper folding.
   1. Molecular chaperones may protect protein structure during heat shock.
   2. The heat shock response involves synthesis of heat shock proteins that prevent denaturation of existing proteins.
   3. Heat shock proteins and other chaperones prevent aggregation of denatured or newly synthesized proteins.
   4. Chaperones also move newly synthesized proteins across membranes.
   5. The protein GroEL is synthesized in *E. coli* and is essential for the proper folding of other cellular proteins.

Key Figure

Figure 2.11. An overview of the types of biological molecules that make up various cellular structures.

Questions for Thought:

1. Of the four kinds of biological macromolecules, which are capable of encoding information? What feature of a molecule makes it able to do so?

2. Of the four kinds of biological macromolecules, which ones can act as catalysts? Which one is responsible for most cellular catalysis?

3. It has been said that the chromatin of a single human cell contains as much information as several million encyclopedias, whereas the information content of glycogen or cellulose is equivalent to that of wallpaper. Explain.

4. Imagine ways in which each of the four types of macromolecules ends up in the proper cellular compartment. As you study cell biology, the mechanisms by which this happens will be further explained.

Review Problems

1. Give the significance (not the definition) of the following terms or phrases. Say what they do or why they are important. For example:
    water: *the fluid matrix of the cell; water is a solvent for the polar molecules of the cell and determines the structure of biological molecules and their interactions.*

    a. electronegativity

    b. free radicals

    c. carbon

    d. functional groups

    e. ß(1 →4) glycosidic linkages in cellulose

    f. amphipathic nature of phospholipids

    g. protein primary structure

    h. site-directed mutagenesis

    i. Anfinsen's experiments with ribonuclease

    j. molecular chaperones

2. Compare and contrast the following:

   a. protein secondary (2°) structure vs. tertiary (3°) structure

   b. glycogen vs. starch

   c. saturated vs. unsaturated hydrocarbons

   d. fat vs. oil

   e. purines vs. pyrimidines

Multiple Choice

1. The element neon (Ne) has eight electrons in its outermost electron shell. How many covalent bonds will Ne readily form?
   a. none
   b. one
   c. two
   d. four

2. Of the following elements, which is likely to form the least polar covalent bonds with hydrogen:
   a. nitrogen
   b. oxygen
   c. carbon
   d. phosphorus

3. Of the following chemical bonds, which is the strongest in water?
   a. $HOH_2C–CH_2OH$
   b. $Na–Cl$
   c. $CH_3COO–H$
   d. $HO–H$

4. Which of the following characteristics does not apply to water?
    a. The water molecule is asymmetric.
    b. The water molecule readily forms hydrophobic interactions.
    c. The covalent bonds in water are highly polarized.
    d. All three atoms in the water molecule readily form hydrogen bonds.

5. The major cellular macromolecules include:
    a. proteins, lipids, amino acids, and carbohydrates.
    b. proteins, amino acids, lipids, and nucleic acids.
    c. proteins, nucleic acids, carbohydrates, and antibodies.
    d. proteins, nucleic acids, carbohydrates, and lipids.

6. If you were to take a triacylglycerol molecule and chemically break the ester bonds, you would end up with two different kinds of molecules. What is the name of the kind of molecule that would be soluble in water?
    a. fatty acid
    b. steroid
    c. glycogen
    d. glycerol

7. Which of the following has been used as evidence that primitive life forms lacked both DNA and enzymes?
    a. RNA can both code genetic information and act as a catalyst.
    b. DNA and enzymes are only present in the most advanced cells.
    c. Advanced cells lack RNA.
    d. All of the above.

8. Of the following functional groups, which would make a carbon-based compound the least polar?
    a. phosphate
    b. methyl
    c. carboxyl
    d. amino

9. Cytoplasm has a pH of about 7. What is the concentration, in moles/liter, of the hydrogen ion?
    a. 7
    b. 7×10^{-7}
    c. 1×10^{7}
    d. 1×10^{-7}

10. Which of the following acid/base pairs act as natural buffers in living systems?
    a. H_2CO_3/HCO_3^-
    b. $H_2PO_4^-/HPO_4^{2-}$
    c. histidine$^+$/histidine
    d. all of the above

11. What types of isomers are these?

    a. A is a cis isomer and B is a trans isomer.
    b. A is a trans isomer and B is a cis isomer.
    c. A is a D isomer and B is an L isomer.
    d. A is an L isomer and B is a D isomer.

12. Which of the following pairs of functional groups might form weak bonds with one another?
    a. methyl and phosphate
    b. phosphate and carboxyl
    c. carboxyl and amino
    d. carbonyl and methyl

13. If you could grab the ends of each of the following polysaccharides and pull, which one might you be able to stretch the most?
    a. glycogen
    b. starch
    c. cellulose
    d. none of the above

14. What two features of the structure of cholesterol make it somewhat amphipathic?
    a. The single hydroxyl along with the hydrocarbon nature of the remainder of the molecule
    b. The long, branched hydrocarbon along with the four hydrocarbon rings
    c. The five methyl groups along with the hydrocarbon chain
    d. The three six-membered rings along with the single five-membered ring

15. Proteins and macromolecular structures take on their higher order structures
    a. by self-assembly.
    b. with the help of molecular chaperones.
    c. with the help of precursor sequences that are removed from the final structure.
    d. all of the above.

16. The different orders of protein structure are determined by all of the following bond types except:
    a. peptide bonds.
    b. phosphodiester bonds.
    c. disulfide bridges.
    d. hydrogen bonds.

17. The higher order structure of DNA shows symmetry, whereas the higher order structures of most proteins do not. Why isn't protein shape more regular like DNA?
    a. DNA has one main function in cells whereas proteins have many.
    b. The many different amino acid R groups on proteins confer many different shapes.
    c. Some amino acids cause proteins to bend; others cause proteins to flatten.
    d. All of the above.

18. Which amino acids would most likely reside in the membrane-anchoring domain of a membrane-embedded protein?
    a. isoleucine, valine, and phenylalanine
    b. phenylalanine, valine, and aspartate
    c. leucine, threonine, and lysine
    d. lysine, arginine, and histidine

19. ATP is from which general category of biomolecules?
    a. polysaccharides
    b. proteins
    c. nucleotides
    d. amino acids

20. Which of these structures would be least likely to occur in cells?

      a.   b.   c.   d.

Problems and Essays

1. The energy of a mole of photons of light in the electromagnetic spectrum decreases logarithmically as a function of increasing wavelength. In the visible range, light has between 40 and 80 kcal/mole of photons. Below about 400 nm wavelengths, the energy goes up drastically. Using what you know about the energy of chemical bonds, make an argument for the use of sun screen with UV block on a sunny day.

2. Water has been described as an excellent thermal buffer for cells. What does that mean, and what accounts for this property of water?

3. Indicate whether these characteristics describe a protein motif or a protein domain, and why:
   a. Proteins that bind calcium often have a helix-loop-helix structure with hydrophilic amino acids at critical positions.

   b. Several proteins, including tissue plasminogen activator (tPA) and the digestive enzyme chymotrypsin, have a 200 to 300 amino acid region that catalyzes the breakdown of protein. This region is attached to the rest of the protein by an amino acid hinge.

c. Proteins that reside in membranes have long sequences of hydrophobic amino acids that anchor the proteins in the lipid membrane.

d. Proteins that bind DNA are characterized by a zinc finger: a combination of one α helix and two ß strands stabilized with a zinc atom. This structure fits into the groove of DNA.

4. For the following questions, refer to these chemical structures:

a. Which molecule is amphoteric?

b. Which molecules are hydrophobic?

c. Which molecule is amphipathic?

d. Which molecule is a building block of carbohydrates?

e. Which molecules have asymmetric carbons?

f. Which molecules have free rotation around all their carbon-to-carbon bonds?

g. Name as many of these molecules as you can, and tell what general category of compounds each one belongs to.

5. Draw the structure of a tripeptide that would partition itself in the core of the cell membrane.

6. When nucleic acids pair in forming the double helix of DNA, the guanine-cytosine pairing results from the formation of three hydrogen bonds between the purine and the pyrimidine, and the adenine-thymine pairing results from the formation of two hydrogen bonds.

a. Estimate the bond energy present in a mole of guanine-cytosine pairs; a mole of adenine-thymine pairs.

b. Which would be more stable: a strand of DNA rich in G–C pairs, or one rich in A–T pairs?

7. Alpha-helical regions of proteins can form amphipathic cylinders, where one side of the cylinder contains hydrophobic amino acids and the other side contains hydrophilic amino acids. In other words, the hydrophobic and hydrophilic amino acid residues are segregated on opposite sides of the alpha helix. If the hydrophilic and hydrophobic amino acids are intermixed, however, the helix is probably not amphipathic. The primary structures of three different 17-amino acid proteins (A, B, and C) are listed below. The numbers at the head of the table correspond to the position of each amino acid in its respective protein. Only two of these proteins will form an alpha helix; the other will not.

Position	1	2	3	4	5	6	7	8	9	10	11	12	13	14	15	16	17
A.	arg	leu	phe	asp	ser	glu	val	leu	lys	ile	ala	val	ile	arg	phe	leu	leu
B.	ser	leu	ile	pro	val	lys	ile	glu	met	val	pro	pro	asp	phe	val	pro	glu
C.	phe	leu	ile	arg	val	leu	arg	lys	val	phe	arg	val	leu	thr	arg	ile	leu

a. Which protein will not form an alpha helix, and why?

b. Of the helical proteins, which will form an amphipathic helix? Explain your answer.

Helix wheel projection

(HINT: The helix wheel projection diagram shown above takes into account the angles of the bonds within the amino acids in an alpha helix, showing the positions of each amino acid in the final alpha helix. The numbers correspond to the numbered amino acids in the table. Use Figure 2.26 in your book to determine which amino acids are hydrophobic in the three different proteins.)

Web Explorations

Most naturally occurring unsaturated fatty acids have double bonds in the *cis* configuration; in other words, the hydrogen atoms on the carbons that form the double bond emerge from the same side of the molecule. This means that the two larger hydrocarbon chains both emerge from the opposite side, introducing a kink into the otherwise linear structure of the hydrocarbon chain. Remind yourself of the difference between *cis* and *trans* configurations by looking at the diagram on page 49 of the textbook.

Until the 1980s, *trans* fatty acids—those in which the two hydrocarbon chains emerge from opposite side of the fatty acid—were quite rare in animal fat, including human fats. In the past fifteen years, however, the occurrence of *trans* fatty acids in human tissues has steadily climbed. Using the World Wide Web, learn more about the sources of *trans* fatty acids. Why are Americans suddenly consuming as much as five grams of *trans* fatty acids each day? What possible effects might this new molecular structure have on our health? How can you avoid eating these compounds, and why should you try?

3

Bioenergetics, Enzymes, and Metabolism

Learning Objectives

When you have finished this chapter, you should be able to:

1. Define energy and distinguish between different forms of energy.
2. Define thermodynamics and understand the implications of the first and second laws.
3. Understand the concepts of free energy and changes in free energy.
   a. Know the relationship between free energy and thermodynamics, and between free energy and equilibrium.
   b. Compare free energy in different chemical reactions and understand standard free energy, $DG°'$.
   c. Know how free energy determines which direction reactions will proceed in cells.
4. Know the properties of enzymes, including how they affect the rates of reactions in cells, and what factors influence those rates.
   a. Describe the mechanisms of enzyme catalysis.
   b. Describe features of the active site.
   c. Define the kinetic constants, K_M and V_{max}, how they are measured, and what insights they can provide about enzymatically catalyzed reactions.
   d. Give examples of how enzymes are regulated in cells.
5. Define metabolism, and distinguish between anabolism and catabolism.
6. Recognize redox reactions in cellular metabolism, and the common coenzymes that are associated with oxidation and reduction.
7. Give an overview of the anaerobic pathways for metabolizing glucose. Distinguish between glucose catabolism and glucose anabolism.

Key Terms and Phrases

energy
potential energy
kinetic energy
total energy (E and DE)
free energy (DG)
standard free energy (DG°')
endergonic
exergonic
equilibrium constant (K_{eq})
coupled reactions
induced fit
initial velocity
turnover number
noncompetitive inhibitor
metabolic intermediate

thermodynamics
first law of thermodynamics
second law of thermodynamics
spontaneous reaction
entropy (S and DS)
enthalpy (H and DH)
TDS
equilibrium
steady state
thermodynamic stability
enzyme kinetics
maximal velocity (V_{max})
irreversible inhibitor
metabolism
end product

enzyme
catalyst
ribozyme
cofactor
coenzyme
kinetic stability
substrate
activation energy (E_a)
transition state
ES complex
active site
Michaelis constant (K_M)
competitive inhibitor
metabolic pathway
catabolism

anabolism	ATP	NAD+/NADH
NADP+/NADPH	redox reactions	oxidized
reduced	reducing agent	oxidizing agent
dehydrogenase	substrate level phosphorylation	transfer potential
fermentation	reducing power	transhydrogenase
phosphorylase	protein kinase	allosteric modulation
feedback inhibition	glycolysis	gluconeogenesis

Reviewing the Chapter

I. Introduction
   A. Function is closely related to structure, in enzymes and at other levels of organization.
   B. Enzymes speed up biological reactions.
   C. Thermodynamic analysis reveals which processes occur spontaneously.
   D. Chemical reactions in cells are linked sequentially to form metabolic pathways.

II. Bioenergetics
   A. Cells require energy to survive.
   B. Energy is the capacity to do work.
      1. Potential energy is the potential to do work.
      2. Kinetic energy is in the process of doing work.
   C. Thermodynamics is the study of changes in energy.
      1. The first law of thermodynamics:
         a. Energy cannot be created or destroyed.
         b. Energy can be transduced or converted from one form to another.
            (1) Cells are capable of energy transformation.
            (2) Photosynthesis is the transduction of light energy into chemical energy.
            (3) All cells rely, directly or indirectly, on photosynthesis for energy.
         c. The universe can be divided into system and surroundings.
            (1) The system is a subset of the universe under study.
            (2) The surroundings are everything in the universe that is not part of the system.
         d. Systems can be closed or open.
            (1) Closed systems exchange energy with the surroundings.
            (2) Open systems exchange matter and energy with the surroundings.
         e. When there is an energy transduction (ΔE) in a system, heat content may increase or decrease, and work may be done.
         f. Total energy changes (ΔE) are independent of pathway.
         g. The first law does not predict whether an energy change will be positive or negative.
      2. The second law of thermodynamics:
         a. In any energy transformation, the entropy of the universe is increased.
            (1) Entropy is a measure of randomness or disorder.
            (2) Entropy is energy not available to do additional work.
         b. Living systems maintain a state of order, or low entropy.
            (1) Order in living systems occurs at the expense of order in the environment.

(2) Order in living systems requires a constant input of energy.
3. Free energy:
   a. The first and second laws can be combined and expressed mathematically.
      (1) In a change of state, total energy (ΔH) has an entropy component (S) and a free energy component (ΔG).
      (2) This can be expressed as $\Delta H = \Delta G + T\Delta S$.
      (3) Free energy, ΔG, represents energy available to do work.
      (4) ΔG is a measure of spontaneity of a reaction.
         (a) $\Delta G < 0$, transformation is spontaneous (exergonic).
         (b) $\Delta G > 0$, transformation requires energy input (endergonic).
      (5) The spontaneity of a reaction depends on both the enthalpy, ΔH, and the entropy, ΔS.
   b. Free-Energy changes in chemical reactions:
      (1) All chemical reactions are theoretically reversible.
      (2) All chemical reactions spontaneously proceed toward equilibrium.
      (3) The rates of chemical reactions are proportional to the concentrations of reactants.
      (4) At equilibrium, the free energies of the products and reactants are equal, or $\Delta G = 0$.
   c. Free energy changes of reactions are compared under standard conditions.
      (1) Standard free energy changes, $\Delta G°'$, are described for each reaction under specific, standard, conditions.
      (2) Standard conditions do not typify cellular conditions, but are useful for making comparisons.
      (3) Standard free energy changes are related to equilibrium: $\Delta G°' = -RT \ln K'eq$.
   d. Free energy changes of reactions under non-standard conditions are calculated by correcting $\Delta G°'$ for prevailing conditions: $\Delta G = \Delta G°' + RT \ln [C][D]/[A][B]$.
      (1) Prevailing conditions may cause ΔG to be negative, even when $G°'$ is positive.
      (2) Making ΔG negative may involve coupling endergonic and exergonic reactions in a sequence.
      (3) Making ΔG negative may involve coupling endergonic and exergonic reactions simultaneously.
         (a) Simultaneously coupled reactions have a common intermediate.
         (b) ATP hydrolysis is often coupled to endergonic reactions in cells.
4. Equilibrium versus steady-state metabolism.
   a. Cellular metabolism is nonequilibrium metabolism.
   b. Cells are open thermodynamic systems.
   c. Cellular metabolism exists in a steady state.
      (1) Concentrations of reactants and products remain constant, but not at equilibrium.
      (2) New substrates enter; products are removed.
      (3) Maintaining a steady state requires a constant input of energy, whereas maintaining equilibrium does not.

III. Enzymes as Biological Catalysts
   A. Enzymes speed up chemical reactions.

1. Enzymes can function when removed from cells, as discovered in 1897 by Hans and Eduard Büchner.
2. Enzymes are almost always proteins, as discovered in 1926 by James Sumner.
3. Enzymes may be conjugated with nonprotein components.
   a. Cofactors are inorganic enzyme conjugates.
   b. Coenzymes are organic enzyme conjugates.

B. Properties of Enzymes:
   1. Enzymes are present in cells in small amounts.
   2. Enzymes are not permanently altered during the course of a reaction.
   3. Enzymes cannot affect thermodynamics of reactions, only rates.
   4. Enzymes may increase the rate of a reaction one trillion (10^{12}) fold.
   5. Enzymes are highly specific for their particular reactants, called substrates.
   6. Enzymes produce only appropriate metabolic products.
   7. Enzymes can be regulated to meet the needs of a cell.

C. Overcoming the activation energy barrier:
   1. A small energy input, the activation energy, is required to initiate even favorable (exergonic) reactions.
      a. The activation energy barrier slows the progress of thermodynamically unstable reactants.
      b. Thermodynamically unstable compounds can be kinetically stable.
      c. Reactant molecules that reach the peak of the activation energy barrier are in the transition state.
   2. Catalysts such as enzymes lower the activation energy.
      a. Without a catalyst, only a few substrate molecules reach the transition state.
      b. With a catalyst, a large proportion of substrate molecules can reach the transition state.

D. The active site and molecular specificity:
   1. An enzyme interacts with its substrate to form enzyme-substrate (ES) complex.
   2. The substrate binds to a portion of the enzyme called the active site.
   3. The active site and the substrate have complementary shapes accounting for substrate specificity.

E. Mechanisms of enzyme catalysis:
   1. Substrate orientation means enzymes hold substrates in the optimal position for the reaction.
   2. Changes in the reactivity of the substrate temporarily stabilize the transition state.
      a. Acidic or basic R groups on the enzyme may change the charge of the substrate.
      b. Charged R groups on the enzyme may attack, bend, or link with the substrate.
      c. Cofactors of the enzyme increase the reactivity of the substrate by removing or donating electrons.
   3. Inducing strain in the substrate
      a. Shifts in conformation after binding cause an induced fit between enzyme and substrate.
      b. Covalent bonds of the substrate are strained.
   4. Conformational changes and catalytic intermediates
      a. Various changes in atomic and electronic structure occur in both the enzyme and substrate during a reaction.
      b. Researchers using time-resolved crystallography have determined the three dimensional structure of an enzyme at successive stages during a single catalyzed reaction.

F. Enzyme kinetics:
1. Kinetics is study of rates of enzymatic reactions under various experimental conditions.
2. The rates of enzymatic reactions increase with increasing substrate concentrations until the enzyme is saturated.
    a. At saturation every enzyme molecule is working at maximum capacity.
    b. The velocity at saturation is called maximal velocity, V_{max}.
    c. The turnover number is the number of substrate molecules converted to product per minute per enzyme molecule at V_{max}.
3. The Michaelis constant (K_M) is the substrate concentration at one-half of V_{max}.
    a. The units of K_M are concentration units.
    b. The K_M may reflect the affinity of the enzyme for the substrate.
4. Temperature and pH can greatly influence enzymatic reaction rates.
5. Plots of the inverses of velocity versus substrate concentrations, such as the Lineweaver-Burk plot, facilitate estimating K_M and V_{max}.

G. Enzyme inhibitors slow the rates of enzymatic reactions.
1. Irreversible inhibitors bind tightly to the enzyme.
2. Reversible inhibitors bind loosely to the enzyme.
    a. Competitive inhibitors:
        (1) compete with the enzyme for active sites.
        (2) usually resemble the substrate in structure.
        (3) can be overcome with high substrate/inhibitor ratios.
    b. Noncompetitive inhibitors:
        (1) bind to sites other than active sites and inactivate the enzyme.
        (2) the maximum velocity of the population of enzyme molecules cannot be reached.
        (3) cannot be overcome with high substrate/inhibitor ratios.

IV. **The Human Perspective: The Growing Problem of Antibiotic Resistance**
A. Antibiotics target bacterial metabolism without harming the human host.
1. Enzymes involved in the synthesis of the bacterial cell wall.
2. Components of the system by which bacteria duplicate, transcribe, and translate their genetic information.
3. Enzymes that catalyze metabolic reactions that occur specifically in bacteria.
B. Antibiotics have been misused with dire consequences:
1. Susceptible cells are killed, leaving rare and resistant cells to survive and repopulate.
2. Bacteria become resistant to antibiotics by acquiring genes from other bacteria by transformation, transduction, and conjugation.

V. **Metabolism**
A. Metabolism is the collection of biochemical reactions that occur within a cell.
B. Metabolic pathways are subsets of metabolism.
1. A metabolic pathway is a sequence of chemical reactions.
2. Each reaction in the sequence is catalyzed by a specific enzyme.
3. Pathways are usually confined to specific cellular locations.
4. Enzymes of a given pathway may be physically linked to one another.
5. Pathways convert substrates into end products via a series of metabolic intermediates.
C. An overview of metabolism:
1. Catabolic pathways break down complex substrates into simple end products.
    a. They provide raw materials for the cell.

b. They provide chemical energy for the cell.
  2. Anabolic pathways synthesize complex end products from simple substrates.
     a. They require energy.
     b. They use ATP and NADPH from catabolic pathways.
  3. Anabolic and catabolic pathways are interconnected.
     a. In stage I, macromolecules are hydrolyzed into their building blocks.
     b. In stage II, building block molecules are further degraded into a few common metabolites of small molecular weight.
     c. In stage III, small molecular weight metabolites like acetyl CoA are further degraded, yielding ATP.
  4. Many metabolic pathways are common to all cells; they arose early in the course of evolution.
D. Oxidation and reduction– a matter of electrons:
  1. Oxidation-reduction (redox) reactions involve a change in the electronic state of reactants.
     a. When a substrate gains electrons, it is reduced.
     b. When a substrate loses electrons, it is oxidized.
     c. When one substance gains (or loses) electrons, some other substance must donate (or accept) those electrons; redox substrates must occur in pairs.
        (1) In a redox pair, the substrate that gains electrons is an oxidizing agent.
        (2) The substrate that donates electrons is a reducing agent.
  2. Redox reactions with organic substrates involve atoms that are bound to carbon.
E. The capture and utilization of energy:
  1. Reduced atoms can be oxidized, releasing energy to do work.
  2. The more a substance is reduced, the more energy can be released.
     a. Fats and carbohydrates are highly reduced.
     b. Fats and carbohydrates are oxidized in cells to release energy to do work.
     c. Glucose is a carbohydrate central to the energy-producing pathways of cells.
F. Glycolysis and ATP formation:
  1. Glycolysis is the first stage in the catabolism of glucose.
     a. It occurs in the soluble portion of the cytoplasm.
     b. It begins with glucose and results in the formation of pyruvate.
  2. Of the reactions of the glycolytic pathway, all but three are near equilibrium ($\Delta G = 0$) under cellular conditions.
  3. The driving forces of the glycolytic pathway are the three reactions that are far from equilibrium.
  4. There are 10 reactions of glycolysis:
     a. Glucose is phosphorylated to glucose 6-phosphate at the expense of ATP.
     b. Glucose-6-phosphate is isomerized to fructose-6-phosphate.
     c. Fructose-6-phosphate is phosphorylated to fructose 1,6-bisphosphate using another ATP.
     d. Fructose 1,6-bisphosphate is split into two three-carbon phosphorylated compounds.
     e. NAD^+ is reduced to NADH when glyceraldehyde 3-phosphate is converted to 1,3-bisphosphoglycerate.
        (1) Dehydrogenase enzymes oxidize and reduce cofactors.
        (2) NAD^+ is a nonprotein cofactor loosely associated with glyceraldehyde phosphate dehydrogenase.

                    (3) NAD⁺ can undergo oxidation and reduction at different places in the cell.
                    (4) NADH donates electrons to the electron transport chain in the mitochondria.
                    (5) Energy from the redox reaction of the electron transport chain is used to make ATP via oxidative phosphorylation.
            f. ATP is formed when 1,3 bisphosphoglycerate is converted to 3-phosphoglycerate by 3-phosphoglycerate kinase.
                    (1) Kinase enzymes transfer phosphate groups.
                    (2) Substrate-level phosphorylation occurs when ATP is formed by a kinase enzyme.
                    (3) ATP formation is only moderately endergonic compared with other phosphate transfer reactions in cells.
            g. 3-phosphoglycerate is converted to pyruvate via three sequential reactions, one of them a kinase that phosphorylates ADP.
        5. Glycolysis can generate a net of two ATPs for each glucose.
        6. Glycolysis occurs in the absence of oxygen; i.e., it is anaerobic.
        7. The end product, pyruvate, can enter anaerobic or aerobic catabolic pathways.
    G. Anaerobic oxidation of pyruvate: The process of fermentation:
        1. Fermentation restores NAD⁺ from NADH.
            a. Under anaerobic conditions, glycolysis depletes the supply of NAD⁺ by reducing it to NADH.
            b. In fermentation, NADH is oxidized to NAD⁺ by reducing pyruvate.
                    (1) In muscle cells and tumor cells, pyruvate is reduced to lactate.
                    (2) In yeast and other microbes, pyruvate is reduced and converted to ethanol– the basis of the alcoholic beverage industry.
        2. Glycolysis followed by fermentation is inefficient: only about 10% of the energy of glucose is captured in ATP.
    H. Reducing power:
        1. Anabolic pathways require a source of electrons to form large molecules.
        2. NADPH donates electrons to form large biomolecules.
            a. NADPH is a nonprotein cofactor similar to NADH.
            b. The supply of NADPH represents the cell's reducing power.
            c. NADP⁺ is formed by phosphate transfer from ATP to NAD⁺.
        3. NADPH and NADH are interconvertible, but have different metabolic roles.
            a. NADPH is oxidized in anabolic pathways.
            b. NAD⁺ is reduced in catabolic pathways.
            c. The enzyme transhydrogenase catalyzes the transfer of hydrogen atoms from one cofactor to the other.
                    (1) NADPH is favored when energy is abundant.
                    (2) NADH is used to make ATP when energy is scarce.
    I. Metabolic regulation:
        1. Cellular activity is regulated according to need.
        2. Regulation may involve controlling key enzymes of metabolic pathways.
        3. Enzymes are controlled by alterations in active sites.
            a. Covalent modification:
                    (1) Enzymes are often regulated by phosphorylation:
                            (a) Protein kinases transfer a phosphate group from ATP to the enzyme being regulated.
                            (b) Phosphate groups are added to tyrosine residues by one type of protein kinase.

(c) Phosphate groups are added to serine or threonine
residues by a second type of protein kinase.
(2) Activation of phosphorylase by phosphorylase kinase was first
described in the 1950s by Fischer and Krebs.
b. Allosteric modulation:
(1) Enzymes can be regulated by compounds binding to allosteric
sites.
(2) In feedback inhibition, the product of the pathway
allosterically inhibits one of the first enzymes of the pathway.
J.   Separating catabolic and anabolic pathways:
1.   Synthesis of biomolecules does not proceed via the same reactions as
breakdown of those molecules, although the two pathways may have steps
in common.
a. Some catabolic steps are essentially irreversible due to large G values.
b. Irreversible steps in catabolic pathways are accomplished by different
mechanisms and different enzymes in anabolic pathways.
2.   Glycolysis and gluconeogenesis are the catabolic and anabolic pathways of
glucose metabolism.
a. Synthesis of fructose 1,6-bisphosphate is coupled to hydrolysis of ATP
by phosphofructokinase in glycolysis.
b. Breakdown of fructose 1,6-bisphosphate is via simple hydrolysis by
fructose 1,6-bisphosphatase in gluconeogenesis.
c. Phosphofructokinase is regulated by feedback inhibition with ATP as
the allosteric inhibitor.
d. Fructose 1,6-bisphosphatase is regulated by covalent modification
using phosphate binding.
e. ATP levels are highly regulated.

VI. **Experimental Pathways: Determining the Mechanism of Lysozyme Action**
A. Lysozyme breaks glycosidic bonds of bacterial cell walls.
B. The mechanism of action was theorized in 1966 by David Phillips.
1.   The mechanism was first modeled using x-ray crystallography information.
2.   Predictions based on the model were tested and found to support the model.
3.   The enzyme and the substrate fit together best while in the transition state.
4.   The best inhibitors mimic the substrate in its transition state and are called
TSAs, or transition state analogs.
C. Recent work using site-directed mutagenesis has further confirmed the mechanism
proposed by Phillips.

Key Figure

Figure 3.8. Energy of activation and enzymatic reactions.

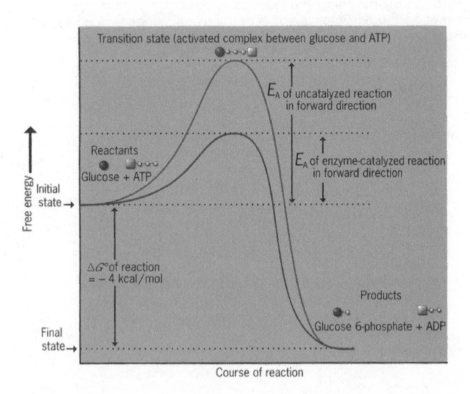

Questions for Thought:

1. How does the activation energy prevent reactions from reaching thermodynamic equilibrium?
2. Give some examples of how enzymes lower the activation energy.
3. Why would it be inappropriate for cells to overcome the energy of activation by raising the mean energy level of all the reactant molecules instead of lowering the energy of activation barrier?
4. Can lowering the activation energy make a reaction go in either direction spontaneously? What if the concentrations of products and reactants were altered, making the reverse reaction a thermodynamically favored reaction. Could the enzyme function to speed up the reverse reaction in those circumstances?

Review Problems

Short Answer

1. Give the significance (not the definition) of the following terms or phrases. Say what they do or why they are important. For example:
    Second law of thermodynamics: *predicts the direction that a reaction will proceed spontaneously.*

    a. standard free energy difference, $\Delta G^{\circ\prime}$

    b. activation energy

    c. kinetic stability

    d. structural homology between a substrate and an enzyme inhibitor

    e. transfer potential

    f. enzyme specificity

    g. reciprocal plots of enzyme velocity vs. substrate concentration

    h. fermentation

    i. reducing power

    j. feedback inhibition

2. Compare and contrast the following:

    a. equilibrium vs. steady state

    b. kinetic stability vs. thermodynamic stability

    c. theoretically reversible reactions vs. essentially irreversible reactions

    d. NAD^+/NADH vs. $NADP^+$/NADPH

    e. competitive vs. noncompetitive inhibition

Multiple Choice

1. Water behind a dam is an example of
    a. kinetic energy.
    b. potential energy.
    c. free energy.
    d. entropy.

2. For the following questions, answer either **I**-increased, **D**-decreased or **U**-unchanged:
    a. When sugar is dissolved in coffee, the entropy of the sugar is_____.
    b. When sea water is evaporated leaving behind salt crystals, the entropy of the water is_

    _____.
    c. The entropy of the salt crystals in (b) above is_____.
    d. When a fertilized egg cell develops into an embryo, the entropy of the living system is_

    _____.
    e. The entropy of the universe in (d) above is_____.

3. For a reaction with a standard free energy change, $\Delta G'^\circ$ = +0.7 kcal/mole, which of the
following statements is true?
    a. The reaction is exergonic.
    b. The reaction would never occur spontaneously.
    c. The reaction could be made to occur by altering the concentrations of reactants and
        products.
    d. None of the above.

4. When a chemical reaction is at equilibrium, which of the following statements is true?
   a. The rate of the forward reaction is equal to the rate of the backward reaction.
   b. The concentrations of reactants and products are equal.
   c. The ratio of the concentrations of reactants and products equals 1.0.
   d. The reaction is proceeding in the forward direction at a rate equal to K'_{eq}.

5. Reactions that have positive standard free energy changes ($\Delta G'^\circ > 0$) can be made to occur in cells by:
   a. coupling them with exergonic reactions via a common intermediate.
   b. manipulating the concentrations of products and reactants such that $\Delta G' < 0$.
   c. coupling them to the hydrolysis of ATP.
   d. all of the above.

6. Cellular metabolism is an example of:
   a. equilibrium metabolism.
   b. steady-state metabolism.
   c. a series of reactions occurring under standard conditions.
   d. a series of reactions that defy the second law of thermodynamics.

7. Any system that is not at equilibrium:
   a. is thermodynamically unstable, although it may be kinetically stable.
   b. is kinetically unstable, although it may be thermodynamically stable.
   c. is rushing toward equilibrium at a very rapid rate.
   d. requires an enzyme to reach equilibrium.

8. Protein catalysts:
   a. increase the rates of chemical reactions in cells while remaining unchanged.
   b. are sometimes referred to as ribozymes.
   c. can change the standard free energy of a cellular reaction.
   d. all of the above.

9. Which of the following statements is not true of enzymes?
   a. Enzymes are proteins that bind to specific substrates and increase the velocity of reactions involving those substrates.
   b. Enzymes function by overcoming the activation energy barrier of a reaction.
   c. Enzymes can only make thermodynamically favorable reactions proceed; they cannot make unfavorable reactions occur.
   d. Enzymes only function when they are in intact cells.

10. As opposed to an uncatalyzed reaction, in an enzymatically catalyzed reaction:
    a. the activation energy of the reaction is lowered, and a larger proportion of substrate molecules has sufficient energy to overcome it.
    b. the activation energy of the reaction remains unchanged, but the substrate molecules are highly energized.
    c. the activation energy of the reaction is increased, thus decreasing the likelihood that any substrate molecules will overcome it.
    d. the activation energy of the reaction is lowered, but fewer substrate molecules can overcome it.

11. When an enzyme is functioning at V_{max}, the rate of the reaction is limited by:
    a. the number of collisions between enzyme and substrate.
    b. the number of substrate molecules in the reaction.
    c. the concentration of the substrate.
    d. the rate at which the enzyme can convert substrate to product.

12. The K_M of a reaction is:
    a. a measure of the affinity of the enzyme for the substrate under some conditions.
    b. a measure of the substrate concentration at exactly one-half of V_{max}.
    c. a kinetic value that can be determined from Lineweaver-Burk plots of enzyme velocities.
    d. all of the above.

13. A certain inhibitor of an enzymatic reaction was determined to act as a competitive inhibitor. Which one of following statements about the inhibitor is most probably true?
    a. The inhibitor acts at a site other than the substrate binding site.
    b. The inhibitor increases K_M but has no effect on V_{max}.
    c. The inhibitor decreases K_M but has no effect on V_{max}.
    d. The inhibitor lowers the maximum velocity of the enzyme.

14. During catabolism, only about 40% of the energy available from oxidizing glucose is used to synthesis ATP. What becomes of the other 60%?
    a. It is lost as heat.
    b. It is used to reduce NADP.
    c. It remains in the products of metabolism.
    d. It is stored as fat.

15. During glycolysis, a net of how many ATPs are produced for each glucose that enters the pathway?
    a. 1
    b. 2
    c. 4
    d. 35

16. For most of the reactions of glycolysis, the G under cellular conditions is close to 0. Why does the pathway continue in the direction of glucose catabolism?
    a. There are three essentially irreversible ($\Delta G \ll 0$) reactions that act as the "driving force" for the pathway.
    b. High levels of ATP keep the pathway going in a forward direction.
    c. The enzymes of glycolysis only function in one direction.
    d. Glycolysis occurs in either direction.

17. Which of the following statements is not true of glycolysis?
    a. ADP is phosphorylated to ATP via substrate-level phosphorylation.
    b. The pathway does not require oxygen.
    c. The pathway oxidizes two moles of NADH to NAD^+ for each mole of glucose that enters.
    d. The pathway requires two moles of ATP to get started catabolizing each mole of glucose.

18. What is the most important function of fermentation from the point of view of the cell?
    a. Fermentation is the basis of the wine, beer, and liquor industries.
    b. Fermentation regenerates NAD^+ so that glycolysis can continue.
    c. Fermentation provides a supply of lactate to metabolizing muscle cells.
    d. Fermentation is an essential step in aerobic metabolism.

19. Which of the following mechanisms are used by cells to regulate metabolism?
    a. Enzymes are regulated by covalent modification.
    b. Enzymes are regulated by allosteric modification.
    c. Metabolic pathways are regulated by compartmentalization.
    d. Cells use all of the above to regulate metabolism.

20. All of the following techniques were used to study the mechanism of lysozyme action except:
    a. x-ray diffraction.
    b. the use of synthetic, chemically altered substrates.
    c. equilibrium thermodynamics.
    d. site directed mutagenesis.

Problems and Essays

1. For the reaction:

    dihydroxyacetone phosphate $\Leftrightarrow$ glyceraldehyde 3-phosphate $\Delta G^{o'} = 1.8$ kcal/mole

    What ratio of product/reactant must be maintained in order to keep the reaction going in a forward direction?

2. For the reaction:

    fumarate $\Leftrightarrow$ malate $K'_{eq} = 02818$

    What is the change in free energy when one mole of fumarate reacts to make one mole of malate under standard conditions?

3. For the reaction:

    fructose 6-phosphate $\Leftrightarrow$ glucose 6-phosphate $\Delta G^{o'} = -0.4$ kcal/mole

    Starting with a 0.50 M solution of fructose 6-phosphate, what is the equilibrium concentration of both fructose 6-phosphate and glucose 6-phosphate?

4. Given the reaction:

    succinate + FAD $\Leftrightarrow$ fumarate + FADH$_2$ $\Delta G^{o'} = 0$ kcal/mole

    a. Starting with a solution of 0.01 M succinate and 0.01 M FAD, how much fumarate will be present at equilibrium?

b. How will adding the enzyme succinic dehydrogenase change your answer to part (a)?

5. For the complete oxidation of glucose:

$$C_6H_{12}O_6 + 6O_2 \Leftrightarrow 6CO_2 + 6H_2O \qquad \qquad \Delta G^{o'} = \text{-686 kcal/mole}$$

Assume that conditions in a cell are such that only 40 percent of the energy of glucose oxidation is captured as ATP, and that concentrations of reactants and products are kept at standard levels.

$$ADP + Pi \Leftrightarrow ATP \qquad \qquad \Delta G^{o'} = \text{+7.3 kcal/mole}$$

How many moles of ATP are formed for each mole of glucose that is oxidized?

6. Palmitic acid is a saturated fatty acid with a chain length of 16 carbons. For the complete oxidation of palmitic acid:

$$COOH(CH_2)_{14}CH_3 + 23\,O_2 \Leftrightarrow 16\,CO_2 + 16H_2O \qquad \qquad \Delta G^{o'} = \text{-686 kcal/mole}$$

Assume that conditions in a cell are such that only 40 percent of the free energy of fatty acid oxidation is capture in the form of ATP. How many moles of ATP are formed for each mole of palmitic acid oxidized?

7. Given the answers to numbers (5) and (6) above, would it be better to eat sugar or fat if you wanted to gain weight? Why? What if you wanted to lose weight?

8. For the following, put either **C** for competitive inhibition, **N** for noncompetitive inhibition, or **B** for both, depending on which type of inhibition the statement best describes.

_____a. The rate of product formation at subsaturating substrate concentrations is adversely affected.

_____b. The inhibitor binds to the active site of the enzyme.

_____c. At very high substrate concentrations, the initial velocity is lower in the presence of the inhibitor.

_____d. The V_{max} is unchanged in the presence of the inhibitor.

_____e. The ES complex can bind the inhibitor.

_____f. In the presence of the inhibitor the Lineweaver-Burk plot is rotated around the y intercept.

9. Given the following data:

	Initial Velocity (nmol product·(min·mg)$^{-1}$)			
[S] (mM)	no inhibitor	inhibitor A	inhibitor B	inhibitor C
0.100	0.400	0.258	0.200	0.333
0.125	0.476	0.316	0.238	0.400
0.167	0.588	0.400	0.294	0.500
0.250	0.769	0.546	0.385	0.667
0.500	1.111	0.857	0.556	1.000
1.000	1.429	1.200	0.714	1.333
2.000	1.667	1.500	0.833	1.600

a. Plot the Lineweaver-Burk graph for the enzymatic reaction without an inhibitor and with one of the three inhibitors, A, B, or C.

b. What is the K_M and V_{max} of the reaction with and without the inhibitor?

c. What kind of inhibitor is it, competitive or noncompetitive? Why?

Web Explorations

Phenylketonuria (PKU) is an inborn error of metabolism characterized by the absence or deficiency of an enzyme, phenylalanine hydroxylase, required to process the essential amino acid phenylalanine. Normal phenylalanine hydroxylase converts phenylalanine to tyrosine, which cells use in protein synthesis. When the enzyme is missing or deficient, abnormally high quantities of the substrate phenylalanine accumulate in the blood plasma.

Using the World Wide Web, investigate the effects of high levels of plasma phenylalanine. How are infants tested for PKU and what treatments are available? How is the disease inherited? Are there specific ethnic groups that have an increased incidence of PKU?

Notes:

4 The Structure and Function of the Plasma Membrane

Learning Objectives

When you have finished this chapter, you should be able to:

1. Visualize the three-dimensional structure of the plasma membrane.
2. Explain the many dynamic roles of the membrane in cellular function.
3. Describe the types of lipids, proteins, and carbohydrates that make up the plasma membrane, how they are arranged in the bilayer, and the forces that hold them together.
4. Explain the relevance of membrane physical state to function. Give examples of techniques that have been used to measure membrane fluidity.
5. Describe the major integral and peripheral proteins of the erythrocyte.
6. Define the ways in which substances can cross membranes. Distinguish between the energetics of transport for charged and uncharged solutes.
7. Characterize the features of the nerve cell that enable it to produce and transmit impulses.
   a. Define the action potential in molecular terms.
   b. Describe how impulses are transmitted between cells.
   c. Give examples of some neurotransmitters and a neurotransmitter receptor.

Key Terms and Phrases

plasma membrane	selective permeability	signal transduction
receptor	lipid bilayer	fluid-mosaic model
amphipathic	phosphoglyceride	head group
sphingolipid	glycolipid	ceramide
cholesterol	leaflet	liposome
integral protein	peripheral protein	lipid-anchored protein
monotopic protein	bitopic protein	polytopic protein
transmembrane protein	freeze-fracture replication	freeze-etching
detergent	transition temperature	phospholipid asymmetry
flippase enzyme	cell fusion	FRAP
membrane domains	cell polarity	band 3
glycophorin A	spectrin	ankyrin
net flux	diffusion	electrochemical gradient
partition coefficient	osmosis	hypertonic
hypotonic	isotonic	turgor
plasmolysis	conductance	voltage-gated channel
ligand-gated channel	facilitated diffusion	facilitative transporter
protein isoforms	active transport	Na^+-K^+ ATPase
red blood cell ghosts	electrogenic pump	P-type pumps
V-type pumps	ABC transporters	cystic fibrosis
CFTR	bacteriorhodopsin	cotransport
symport	antiport	membrane potential
resting potential	depolarization	threshold

51

action potential   refractory period   all-or-none law
nerve impulse   myelin   saltatory conduction
synapse   synaptic cleft   synaptic vesicle
neurotransmitter   acetylcholine receptor

Reviewing the Chapter

I. Introduction
   A. The plasma membrane is the outer boundary of the cell.
   B. Plasma membranes of all cell types exhibit characteristic trilaminar appearance in electron micrographs.

II. An Overview of Membrane Functions
   A. Compartmentalization: Membranes form continuous sheets that enclose intracellular compartments.
   B. Scaffold for biochemical activities: Membranes provide a framework that organizes enzymes for effective interaction.
   C. Selectivity permeable barrier: Membranes allow regulated exchange of substances between compartments.
   D. Transporting solutes: Membrane proteins facilitate the movement of substances between compartments.
   E. Responding to external signals: Membrane receptors transduce signals from outside the cell in response to specific ligands.
   F. Intercellular interaction: Membranes mediate recognition and interaction between adjacent cells.
   G. Energy transduction: Membranes transduce photosynthetic energy, convert chemical energy to ATP, and store energy in ion and solute gradients.

III. A Brief History of Studies on the Structure of the Plasma Membrane
   A. Membranes were found to be mostly composed of lipids because their dissolving power matched that of oil.
   B. The lipid bilayer:
       1. The bilayer accounted for the 2:1 ratio of lipid to cell surface area.
       2. The most energetically favorable orientation for the polar head groups is facing the aqueous compartments outside of the bilayer.
       3. The nature and importance of the lipid bilayer:
           a. Lipid composition can influence the activity of membrane proteins and determine the physical state of the membrane.
           b. The cohesion of bilayers to form a continuous sheet makes cells deformable and facilitates splitting and fusion of membranes.
           c. Amphipathic lipids spontaneously form bilayered sheets or spherical vesicles, called liposomes, in artificial systems.
           d. Liposomes are being tested as vehicles to deliver drugs or DNA molecules within the body.
           e. Spontaneously formed lipid bilayers may have provided the first stable environment for the development of self-replicating molecules.
   C. Protein-lined pores in the membrane account for the movement of polar solutes and ions across cell boundaries.
   D. The fluid-mosaic model is the "central dogma" of membrane biology.
       1. The core lipid bilayer exists in a fluid state, capable of dynamic movement.
       2. Membrane proteins form a mosaic of particles penetrating the lipid to varying degrees.

IV. The Chemical Composition of Membranes
   A. Membrane composition:
      1. The lipid and protein components of membranes are bound together by non-covalent forces.
      2. In addition to lipids and proteins, membranes also contain carbohydrates.
      3. Protein/lipid ratios vary considerably among different membrane types.
   B. Membrane lipids:
      1. Membrane lipids are amphipathic.
      2. There are three main types of membrane lipids:
         a. Phosphoglycerides are diacylglycerides with small functional head groups linked to the glycerol backbone by phosphate ester bonds.
         b. Sphingolipids are ceramides formed by the attachment of sphingosine to fatty acids.
         c. Cholesterol is a smaller and less amphipathic lipid that is only found in animals.
   C. The Nature and Importance of the Lipid Bilayer:
      1. Membrane lipid composition is characteristic of specific membranes.
      2. Lipids give membranes the ability to fuse, form networks, and separate charge.
      3. Lipid bilayers assemble spontaneously in aqueous solutions.
   D. Membrane Carbohydrates:
      1. Membranes contain carbohydrates covalently linked to lipids and proteins on the extracellular surface of the bilayer.
      2. Glycoproteins have short, branched carbohydrates for interactions with other cells and structures outside the cell.
      3. Glycolipids have larger carbohydrate chains that may be cell-to-cell recognition sites.

V. The Structure and Functions of Membrane Proteins
   A. Membrane proteins attach to the bilayer asymmetrically, giving the membrane a distinct "sidedness."
   B. Integral membrane proteins :
      1. Integral proteins are amphipathic, with hydrophobic domains anchoring them in the bilayer and hydrophilic regions forming functional domains outside of the bilayer.
      2. Channel proteins have hydrophilic cores that form aqueous channels in the membrane-spanning region.
   C. Distribution of Integral Proteins: Freeze-Fracture Analysis
      1. Freeze-fracture technique divides the phospholipid leaflets of the membrane.
      2. Integral membrane proteins appear as bumps and pits using the electron microscope.
      3. The heterogeneity of protein distribution differ is apparent.
   D. Studying the structure and properties of integral membrane proteins:
      1. Determining membrane sidedness: The orientation of integral proteins can be determined using nonpenetrating agents that label the proteins.
      2. Identifying transmembrane domains: A string of 20-30 hydrophobic amino acids as determined by hydropathy plots identifies a membrane-spanning domain.
      3. Spatial relationships within an integral membrane protein
         a. Site directed mutagenesis—replacing specific amino acids with others—identifies some spatial relationships.
         b. Electron spin resonance identifies some conformational changes that occur when integral proteins function.

E. Peripheral proteins are attached to the membrane by weak bonds and are easily solubilized.
F. Lipid-anchored membrane proteins are distinguished both by the types of lipid anchor and their orientation.
    1. Glycophosphatidylinositol (GPI)-linked proteins found on the outer leaflet can be released by inositol-specific phospholipases.
    2. Some inner-leaflet proteins are anchored to membrane lipids by long hydrocarbon chains.

VI. Membrane Lipids and Membrane Fluidity

A. Membrane lipids exist in gel or liquid-crystal phases depending on temperature, lipid composition and saturation and the presence of cholesterol.
    1. Liquid-crystal membranes are the norm.
    2. Unsaturated fatty acids lower the temperature at which the liquid-crystal/gel phase transition occurs.
    3. Patches of membrane, called "rafts," may remain gel-like to aid in the function of GPI-anchored proteins that reside there.
B. The importance of membrane fluidity:
    1. The fluidity of membranes is a compromise between structural rigidity and complete fluidity.
    2. Membrane fluidity makes it possible for proteins to move in the membrane and for membranes to assemble and grow.
C. Maintaining membrane fluidity:
    1. Organisms (other than birds and mammals) maintain membrane fluidity as temperature changes by altering the composition of membrane lipids.
    2. Remodeling lipid bilayers involves saturation or desaturation of acyl chains and replacement of acyl chains by lipases or transferases.
    3. The importance of these mechanisms has been verified using mutants unable to carry out certain desaturation reactions in response to cold.
D. The asymmetry of membrane lipids:
    1. The inner and outer membrane leaflets were shown to have different lipid compositions, using lipases that can only access the outer membrane.
    2. Lipid asymmetry gives the membrane leaflets different physical and chemical properties appropriate for the different interactions occurring at the two membrane faces.
    3. Membrane lipids move easily within a leaflet but only rarely "flip-flop" between leaflets.
E. Lipid Rafts:
    1. The outer leaflet of the plasma membrane contains specialized regions or microdomains.
    2. Cholesterol and sphingolipids tend to pack together to form highly ordered microdomains.
    3. Cholestrol-rich microdomains form "lipid rafts" that float within the more fluid and disordered environment.
    4. Lipid rafts provide a favorable environment for cell-surface receptors and GPI-anchored proteins which shown an affinity for such ordered regions of the bilayer.

VII. The Dynamic Nature of the Plasma Membrane

A. The diffusion of membrane proteins after cell fusion:
    1. Cells can be induced to fuse by certain inactivated viruses, or with polyethylene glycol.
    2. Labeled proteins have shown that membrane proteins can move between fused cells.
B. Patterns of protein mobility:

1. Proteins can be labeled and tracked by fluorescence recovery after photobleaching (FRAP) and single particle tracking (SPT).
2. Proteins can be immobile, mobile in a directed manner, or exhibit random movement.
C. Control of membrane motility:
1. Protein movements in membranes are slower than predicted by protein size and membrane viscosity alone.
2. Protein movements are limited by interactions with the cytoskeleton, other proteins, and extracellular materials.
3. Optical tweezers, focused laser beams that can drag tagged proteins within the membrane, indicate that some proteins have barriers to lateral diffusion.
4. Genetically modified proteins missing either intracellular or extracellular domains are less restricted; their rate of lateral diffusion is greater.
D. Membrane Lipid Mobility:
1. Phospholipid diffusion is restricted within the bilayer.
2. Phospholipids are confined for very brief periods to certain areas and then hop from one confined area to another.
3. Fences restricting motion are constructed of rows of integral membrane proteins bound to the membrane skeleton by their cytoplasmic domains.
E. Membrane domains and cell polarity:
1. Differences in protein distribution are evident in cells of organized tissues.
2. In epithelia, the proteins of the apical membrane are distinct from those of the lateral and basal membranes.
F. The plasma membrane of the red blood cell:
1. Highly homogenous preparations of membrane "ghosts" can be prepared inexpensively and simply from erythrocytes, by osmotic lysis.
2. Membrane proteins can be purified and characterized by fractionation using SDS-PAGE electrophoresis.
G. Integral proteins of the erythrocyte membrane:
1. Band 3 is a homodimeric, polytopic glycoprotein that exchanges Cl^- and HCO^{3-} across the red cell membrane.
2. Glycophorin A is a monotopic protein with 16 oligosaccharide chains bearing negative charges that may prevent red cells from clumping.
H. The erythrocyte membrane skeleton:
1. The major component of the internal membrane skeleton of erythrocytes is spectrin.
2. Spectrin molecules are attached to the membrane surface by noncovalent bonds to ankyrin, a peripheral membrane protein which is noncovalently bonded to band 3.
3. Spectrin is linked to other cytoplasmic proteins, including actin and tropomyosin, which maintains the integrity of the membrane.

VIII. **The Movement of Substances Across Cell Membranes**
A. Selective permeability allows for separation and exchange of materials across the plasma membrane.
1. Net flux is the difference between influx and efflux of materials.
2. Flux can occur by passive diffusion and/or active transport.
B. The energetics of solute movement:
1. Diffusion is the spontaneous movement of material from a region of high concentration to a region of low concentration.
2. The free-energy change during diffusion of nonelectrolytes depends on the concentration gradient.
3. The free-energy change during diffusion of electrolytes depends on the electrochemical gradient.

C. Diffusion of substances through membranes:
    1. Diffusion requires both a concentration gradient and membrane permeability.
    2. Lipid permeability is determined by the partition coefficient, molecular size, and polarity.
D. The diffusion of water through membranes:
    1. The diffusion of water through a semipermeable membrane is called osmosis.
    2. Water diffuses from areas of lower solute concentration to areas of higher solute concentration.
    3. Cells swell in hypotonic solutions, shrink in hypertonic solutions and remain unchanged in isotonic solutions.
    4. Plant cells develop turgor in hypotonic solutions because cell walls prevent swelling. In hypertonic solutions the plant cell undergoes plasmolysis.
    5. Aquaporins are specialized protein-lined channels that allow the passive movement of water across membrane.
E. The diffusion of ions through membranes:
    1. Ions cross membranes through ion channels.
    2. Ion channels are selective and bidirectional, allowing diffusion in the direction of the electrochemical gradient.
    3. Superfamilies of ion channels have been discovered by cloning, analysis of protein sequences, site directed mutagenesis, patch clamping experiments.
    4. Ion channels can be open or closed. They are voltage- or ligand-gated.
    5. The voltage-gated potassium channel (called Kv) contains six membrane-spanning helices.
        a. Both N and C termini are cytoplasmic.
        b. A single channel has 4 subunits arranged to create an ion-conducting pore.
        c. The channel can be opened, closed, or inactivated.
        d. The S4 transmembrane helix is voltage sensitive.
        e. The crystal structure of the bacterial postassium channel show that a short amino acid domain acts to select potassium and no other ions.
F. Facilitated diffusion:
    1. Large or hydrophilic substances require facilitation to cross membranes.
    2. Facilitated diffusion is passive, specific, saturable, and regulated.
    3. The glucose transporter:
        a. The gradient for glucose entry into the cell is maintained by phosphorylation of glucose in the cytoplasm.
        b. Insulin stimulates glucose uptake by causing the insertion into the cell membrane of vesicles containing preformed glucose transporters.
        c. Type I diabetes is due to lack of insulin, type II diabetes is due to lack of a response to insulin.
G. Active transport:
    1. Active transport maintains the gradients for potassium, sodium, calcium, and other ions across the cell membrane.
    2. Active transport couples the movement of substances against gradients to ATP hydrolysis.
H. Coupling active transport to ATP hydrolysis:
    1. The Na^+-K^+ ATPase requires K^+ outside, Na^+ and ATP inside, and is inhibited by ouabain.
    2. The ratio of Na^+:K^+ pumped is 3:2.
    3. The ATPase is a P-type pump. Phosphorylation causes changes in conformation and ion affinity that allow transport against gradients.

4. The Na^+-K^+ ATPase is found only in animals and evolved early as a means to regulate volume and create large Na^+ and K^+ gradients.

   I. Other ion transport systems:
1. Other P-type pumps include H^+ and Ca^{2+} ATPases, and H+/K+ ATPases.
2. Vacuolar (V-type) pumps use ATP, but are not phosphorylated during pumping.
3. ATP-binding cassette (ABC) transporters have regulatory ATP-binding sites.

   J. Using light energy for active transport of ions.
1. Some archaebacteria use a protein called bacteriorhodopsin, which absorbs light energy to transport protons out of the cell.
2. The proton gradient is used to make ATP.

   K. Cotransport: Gradients created by active ion pumping store energy that can be coupled to other transport processes.

IX. The Human Perspective: Defects in Ion Channels as a Cause of Inherited Disease

   A. Several inherited disorders have been linked to mutations in genes encoding ion channel proteins.

   B. CF is a genetic disease characterized by abnormal fluid secretions from tissues and caused by a defective chloride channel.

   C. Genetic analysis revealed a deletion in an ABC transporter (the CFTR polypeptide) with two nucleotide-dependent regulatory sites.

   D. A defect prevents normal insertion of the CFTR polypeptide into the membrane.

   E. Cystic fibrosis is a good candidate for gene therapy by inhalation of aerosols.

X. Membrane Potentials and Nerve Impulses

   A. Potential differences exist when charges are separated.
1. Membrane potentials have been measured in all types of cells.
2. Neurons are cells that are specialized for information transmission using changes in membrane potentials.
     a. The dendrites receive incoming information.
     b. The cell body contains the nucleus and metabolic center of the cell.
     c. The axon is a long extension that conducts outgoing electrical information.

   B. The resting potential:
1. The resting potential is the membrane potential of a nerve or muscle cell, subject to dramatic changes when the cell is activated.
2. K^+ gradients maintained by the Na^+-K^+ ATPase are responsible for the resting membrane potential.
3. The Nernst equation is used to calculate the voltage equivalent of the concentration gradient for particular ions.
4. The negative resting membrane potential is near the negative Nernst potential for K^+ and far from the positive Nernst potential for Na^+.

   C. The action potential:
1. When cells are stimulated, Na^+ channels open, causing membrane depolarization.
2. When cells are stimulated to threshold, voltage-gated Na^+ channels open, triggering an action potential.
3. Sodium channels are inactivated (cannot reopen) immediately following an action potential, producing a short refractory period when the membrane cannot be stimulated.
4. Excitable membranes exhibit "all-or-none" behavior.

   D. Propagation of action potentials as an impulse:

1. Action potentials produce local membrane currents depolarizing adjacent regions of the membrane that propagate the action potential.
2. Speed is of the essence: The speed of a neural impulse depends on the axon diameter and whether the axon is myelinated.
    a. Resistance to local current flow decreases as diameter increases.
    b. Myelin sheaths cause saltatory conduction.
E. Neurotransmission: Jumping the synaptic cleft:
    1. Presynaptic neurons communicate with postsynaptic neurons (synapses) or muscles (neuromuscular junctions) across a gap (synaptic cleft).
    2. Chemical transmitters released from the presynaptic cell diffuse to receptors on the postsynaptic cell.
    3. The bound transmitter can depolarize (excite) or hyperpolarize (inhibit) the postsynaptic cell.
    4. Transmitter action is terminated by reuptake or enzymatic breakdown.
    5. Interference with the destruction or reuptake of neurotransmitters can have dramatic physiological and behavioral effects.
    6. Synapses control the directional flow of information in the nervous system.
F. Synaptic Plasticity:
    1. Synapses connecting neurons to their neighbours can become strengthened over time by long term potentiation (LTP).
    2. The NMDA receptor binds to the neurotransmitter glutamate and opens an internal cation channel.
    3. Subsequent influx of Ca^{2+} ions triggers a cascade of biochemical changes that lead to synaptic strengthening.
    4. LTP inhibitors reduce the learning ability of laboratory animals.

XI. **Experimental Pathways: The Acetylcholine Receptor**
A. Claude Bernard discovered that curare paralyzed muscle function without blocking either nerve or muscle impulses.
B. Langley postulated a "chemical transmitter" and "receptive substance" that bound both curare and nicotine.
C. Loewi used two hearts to show that "vagusstoff" (acetylcholine) formed in one heart could stop contraction in the second.
D. The electric fish Torpedo is an excellent source of nicotinic acetylcholine (nACh) receptors and acetylcholinesterase.
E. The snake venom α-bungarotoxin and the nonionic detergent Triton X-100 allowed affinity purification of the nACh receptor.
F. Reconstituting purified receptors into artificial lipids proved that the nACh receptor was a cation channel.
G. The structure of the receptor has been studied by both electron microscopy and genetic methods.
H. A 43K protein is shown to anchor the receptor to the postsynaptic region.

Key Figure

Figure 4.4c A brief history of the structure of the plasma membrane: a current representation.

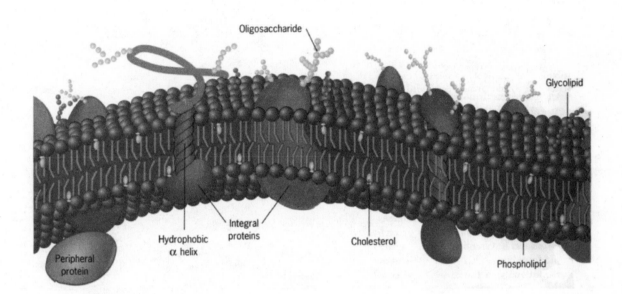

Questions for Thought:

1. The inner and outer leaflets of the cell membrane are different in both composition and function. What leaflet asymmetries are apparent from this cartoon drawing of a membrane? What asymmetries exist that are not apparent in this drawing?
2. In this drawing, the membrane proteins are evenly distributed throughout the lipid matrix. Can you think of a membrane, or portion of a membrane, in which proteins might be clumped together? What functional role might clumps of proteins serve in membranes?
3. Can you identify which proteins in this diagram are peripheral, which are lipid-anchored, and which are integral?
4. What factors influence the frequency with which membrane-bound proteins collide with one another?

Review Problems

1. Give the significance (not the definition) of the following terms or phrases. Say what they do or why they are important. For example:
    amphipathic nature of phospholipids: *results in the spontaneous formation of the bilayer structure characteristic of cell membranes.*

    a. membrane leaflet asymmetry

    b. partition coefficient

    c. flippase enzymes

    d. freeze-fracture and freeze etching

    e. detergents

    f. hydropathy plots

    g. chemical-gated channels

    h. Na^+-K^+ ATPase

    i. membrane potentials

    j. myelin

2. Compare and contrast the following:

   a. the membrane surface facing inside the cell vs. that facing outside

   b. membrane lipid mobility vs. membrane protein mobility

   c. diffusion vs. facilitated diffusion

   d. membrane potential vs. resting potential

   e. myelinated nerve cells vs. large-diameter nerve cells

Multiple Choice

   1. Of the four classes of cellular macromolecules, which one is not a component of cell membranes?
      a. proteins
      b. lipids
      c. nucleic acids
      d. carbohydrates

   2. How do membranes function as a locus of biochemical reactions?
      a. Membranes separate the reactants from the products of enzymatic reactions.
      b. Membranes hold enzymes of sequential reactions in close proximity.
      c. Membrane lipids catalyze important cellular reactions.
      d. Membranes increase the activation energy for catalysis.

   3. Features of the fluid mosaic model of membrane structure include:
      a. a lipid bilayer.
      b. dynamic motion of both membrane lipids and proteins.
      c. proteins that may either rest on the surface or penetrate the membrane.
      d. all of the above.

   4. The relative proportion (by weight) of lipid and protein in a membrane:
      a. is always about equal.
      b. varies depending upon the source of the membrane.
      c. is always characterized by more lipid than protein.
      d. is always characterized by more protein than lipid.

5. Which of the following is not true of membrane phosphoglycerides?
   a. They include phosphatidylcholine, phosphatidylethanolamine and cholesterol, among others.
   b. They are amphipathic.
   c. They have two fatty acids attached to a glycerol molecule, and a polar head group attached via a phosphate.
   d. They can have both saturated and unsaturated fatty acids.

6. Gorter and Grendel's earliest evidence that membranes were formed from a lipid bilayer included:
   a. microscopic examination of red blood cell membranes.
   b. measuring the surface area of lipids extracted from red blood cells, and comparing that to the surface area of the cells.
   c. measuring the mobility rates of membrane proteins.
   d. all of the above.

7. Membrane carbohydrates occur as:
   a. either glycoproteins or glycolipids, but are not free in the membrane.
   b. either glycoproteins, or free in the membrane, but are not attached to lipids.
   c. either glycolipids or free in the membrane, but are not attached to proteins.
   d. none of the above.

8. A membrane-affiliated protein is isolated by detergent extraction, and found to be amphipathic and lacking any nonprotein components. To which class of membrane proteins does it belong?
   a. integral proteins
   b. peripheral proteins
   c. lipid-anchored proteins
   d. none of the above

9. Site-directed cross-linking is a technique that gives insight into what property of integral membrane proteins?
   a. number of transmembrane domains
   b. on which side of the membrane the N and C termini are found
   c. proximity of various membrane-spanning domains to each other
   d. none of the above

10. A membrane is examined using freeze-fracture and freeze-etch analysis. The surfaces are shown to have almost no pits or bumps. Which of the following is the most likely source of the membrane?
    a. mitochondria, inner membrane
    b. myelin
    c. red blood cell
    d. all of the above

11. Which one of the following organisms would you predict has the highest proportion of unsaturated fatty acids in their membranes?
    a. Antarctic fish
    b. cactus
    c. bacteria from thermal hot springs
    d. humans

12. Which of the following statements about the mobility of membrane components is true?
    a. Proteins and lipids move equally rapidly within the plane of the membrane.
    b. Lipids generally move more rapidly than proteins within the plane of the membrane.
    c. Proteins generally move more rapidly than lipids within the plane of the membrane.
    d. Lipids and proteins "flip-flop" between membrane leaflets more readily than they move within the plane of the membrane.

13. Substance X is a charged molecule of small molecular weight. It is found at higher concentrations outside of cells than inside. How might substance X enter cells?
    a. diffusion through a channel
    b. diffusion through the lipid bilayer
    c. active transport
    d. none of the above

14. The distribution of K^+ across an artificial membrane was measured, and the concentrations were found to be equal on both sides. Which of the following statements is true about the distribution of K^+?
    a. K^+ must be at equilibrium across the cell membrane.
    b. K^+ cannot be at equilibrium across the membrane.
    c. There cannot be a membrane potential under these conditions.
    d. More information is needed to determine whether K^+ is at equilibrium.

15. If a freshwater plant were put in sea water, its cells would:
    a. undergo plasmolysis.
    b. lyse.
    c. swell with turgor pressure.
    d. all of the above.

16. The cell in number (15) above would be in an environment that was:
    a. hypertonic.
    b. hypotonic.
    c. isotonic.
    d. hypo-osmotic.

17. Which of the following is an example of an ion transporter?
    a. Na^+-K^+ ATPase
    b. ABC transporter
    c. CFTR
    d. all of the above

18. Arrange the statements below in the proper sequence of events that occur in an action potential:
    I. The resting potential is reestablished.
    II. A stimulus depolarizes the membrane to threshold.
    III. Na^+ gates open and Na^+ rushes into the cell.
    IV. K^+ gates open and K^+ rushes out of the cell.
    a. II, III, IV, I
    b. III, IV, II, I
    c. II, IV, III, I
    d. I, II, IV, III

19. A neurotransmitter that binds to a postsynaptic cell and opens K+ channels:
    a. excites the postsynaptic cell.
    b. inhibits the postsynaptic cell.
    c. depolarizes the postsynaptic cell.
    d. must be acetylcholine.

20. Much of what we know about the nicotinic acetylcholine receptor has come from studies of:
    a. the *Torpedo marmarota* electric organ.
    b. the human heart.
    c. simple, single-celled organisms.
    d. *E. coli* bacteria.

Problems and Essays

1. Your textbook states that compartmentalization in one role of cellular membranes. Based on what you have learned in this and earlier chapters, give two examples of how intermixing of intracellular contents would be harmful to a cell.

2. Some integral membrane enzymes depend upon the lipids in their microenvironment not only for scaffolding, but for enzymatic activity. The Na^+-K^+ ATPase is one example. Using techniques described in chapter 4, design an experiment to test the influence of membrane fluidity on the velocity of the Na^+-K^+ ATPase.

3. Some of these "experimental results" are real, and others have been concocted to test your knowledge. Indicate which results are consistent with our current understanding of the structure of cell membranes (C), and which are not consistent (N).

    a._____Freeze-fracture studies of artificial bilayers show no membrane particles, whereas those of natural membranes show many pits and bumps.
    b._____When lipids are extracted from red blood cell membranes and spread in a monolayer on the surface of water, they occupy an area equal to about twice the surface area of the cells from which they were extracted.
    c._____Treatment of natural membranes with high salt solutions results in complete separation of all the lipid from the protein and carbohydrate components.
    d._____If the membrane of a cell is punctured using a micropipette, the membrane will reseal itself after the pipette is removed.
    e. _____Stains that specifically bind to carbohydrates will stick to the inner surface of a cell membrane, but not the extracellular surface.

4. Indicate whether each of the following proteins is integral (**I**), peripheral (**P**), or lipid-anchored (**LA**).

   a. _____ Cytochrome c is affiliated with the inner mitochondrial membrane where it participates in the transfer of electrons in aerobic respiration. It can be dissociated from the membrane by treatment with 3 M KCl, resulting in a fully active, soluble protein.

   b. _____ Ca^{2+} ATPase is a protein affiliated with the internal membrane system of muscle cells. It can be partially isolated from the membrane by treatment with detergents, but usually retains a few molecules of phospholipid. When the detergent is dialyzed away, the protein forms a cloudy suspension.

   c. _____ Acetylcholinesterase is a protein associated with the synaptic membranes of nerve cells. The enzyme can be separated from the membrane in a fully active form by treatment with phospholipase, but not with strong salt solutions. Analysis of its primary structure reveals no long stretches of hydrophobic amino acids.

   d. _____ Porin is present in the membranes of *E. coli*. Its function is to permit the flow of small molecules across the membrane.

   e. _____ The N-terminus of the bacterial protein bacteriorhodopsin is located on the extracellular surface of the cell, and the C-terminus is on the cytoplasmic side.

   f. _____ Both the N-terminus and the C-terminus of cytochrome b_5 are restricted to the same side of the membrane. The cytochrome remains firmly anchored to the membrane in the presence of strong salt solutions, but can be extracted with detergents. The primary structure reveals one stretch of hydrophobic amino acids.

5. Urea is an uncharged molecule that is produced as waste during the digestion of protein. The kidney has the task of ridding the body of excess urea. The concentration of the urea in the plasma (i.e., on the "inside") is about 0.04 M. The typical urine concentration is 3.00 M. What is the cost, in kcal, of moving a mole of urea out of the body and into the urine?

6. It is possible for the distribution of some substances across a cell membrane to be at equilibrium, even if the concentrations inside and outside the cell are not equal. Explain this.

7. On the planet Loligo, where all living creatures have body temperatures of 298 °K, elements have different symbols than on Earth. The major ions in the nerve cells of the Loligotians and their concentrations (in mM) are as follows:

ion:	Ψ^+	Ξ^+	Θ^-
inside cells	180	6	34
outside cells	21	244	51

Unlike Earthling nerve cells, the resting potential of Loligotian nerve cells is +100 mV, with the inside of the cells positive relative to the outside.

a. Which of the ions is closest to equilibrium in the resting nerve cell?

b. To which of the ions is the Loligotian cell membrane most permeable; in other words, for which ion would you expect to find the most leak channels?

c. Earthling action potential begins by opening membrane channels specific for the ion that is furthest from equilibrium, i.e., voltage-gated Na^+ channels, and allowing such ions to rush toward equilibrium, thus temporarily reversing the resting potential. If the same holds true for the Loligotians, what kind of channels might instigate the action potential on that planet?

Web Explorations

Multiple sclerosis (MS) is a devastating disease that usually affects young adults, often between the ages of 20 and 40. The symptoms vary among individuals, but typically include blurred or double vision, weakness and fatigue, and loss of coordination. In some severe cases, the onset of the disease can involve blindness.

MS occurs when the myelin sheath that surrounds certain neurons begins to disintegrate, also known as demyelination. The causes of demyelinaton are not fully understood, although several theories exist. Using the World Wide Web, learn more about the proposed causes of demyelination in MS. What treatments are currently available and how do they work? Using both your Cell Biology textbook and the web, describe how demyelination causes the symptoms of multiple sclerosis.

5

Aerobic Respiration and the Mitochondrion

Learning Objectives

When you have finished this chapter, you should be able to:

1. Explain the relationship between the structure and function of mitochondria. Relate this to mitochondrial evolution.
2. Summarize the major metabolic pathways that occur in mitochondria.
   a. Understand the central role that the TCA cycle plays in cellular metabolism.
   b. Describe the four types of electron carriers in the electron transport chain, and the four complexes into which they are organized.
3. Describe the transfer of energy in oxidative metabolism, including the forms in which it is captured, stored and transduced.
4. Define the protonmotive force, how it arises, and its electrical and chemical components.
5. Explain how ATP is synthesized, including the machinery, energetics, and regulation.
6. Describe how proteins are targeted for mitochondria.

Key Terms and Phrases

aerobic
outer and inner membranes
intermembrane space
coenzyme A
chemiosmotic mechanism
electron-transfer potential
flavoprotein
ubiquinone (coenzyme Q)
NADH dehydrogenase
cytochrome c oxidase
2,4-dinitrophenol (DNP)
ATP synthase
microbodies
chaperones

anaerobic
cristae
porins
tricarboxylic acid (TCA) cycle
fast-twitch fibers
oxidation-reduction potential
cytochrome
electron transport complex
succinate dehydrogenase
protonmotive force (Δp)
F_1 ATPase
submitochondrial particles
peroxisomes

fermentation
matrix
"high energy" electrons
oxidative phosphorylation
slow twitch fibers
electron-transport chain
prosthetic group
iron-sulfur proteins
cytochrome bc_1
electrochemical gradient
F_0 ATPase
ADP-ATP translocase
glyoxysomes

Reviewing the Chapter

I. Introduction
   A. Oxygen accumulated in the primitive atmosphere after the cyanobacteria appeared.
   B. Aerobes evolved to use oxygen to extract more energy from organic foodstuffs.
   C. In eukaryotic cells, aerobic respiration takes place in the mitochondria.

II. Mitochondrial Structure and Function

  A. Mitochondria have characteristic morphologies despite variability in appearance.
1. Typical mitochondria are bacteria-sized, sausage-shaped organelles but may be round or threadlike.
2. The size and number of mitochondria reflect the energy requirements of the cell.
3. Inner and outer mitochondrial membranes enclose two spaces: the matrix and intermembrane space.
  B. Mitochondrial membranes. Differences between the inner and outer membranes:
1. The outer membrane is less than 50% protein; the inner membrane is more than 75% protein.
2. The inner membrane contains cardiolipin but not cholesterol, both are true of bacterial membranes.
3. The outer membrane contains a large pore-forming protein called porin.
4. The inner membrane is impermeable to even small molecules and ions. The outer membrane is permeable to even some proteins.
  C. The mitochondrial matrix:
1. The matrix contains a circular DNA molecule, ribosomes, and enzymes
2. RNA and proteins can be synthesized in the matrix.
  D. Mitochondria self-replicate by binary fission.

III. Oxidative Metabolism in the Mitochondria

  A. The first steps in oxidative metabolism are carried out in glycolysis.
1. Glycolysis produces pyruvate, NADH, and two molecules of ATP.
2. Aerobic organisms use molecular oxygen to extract more than 30 additional ATPs from pyruvate and NADH.
3. Pyruvate is transported across the inner membrane and decarboxylated to form acetyl CoA, which enters the tricarboxylic acid (TCA) cycle.
  B. The tricarboxylic acid cycle:
1. The TCA cycle is a stepwise cycle where substrate is oxidized and its energy conserved.
2. The two-carbon acetyl group from acetyl CoA is condensed with the four-carbon oxaloacetate to form a six-carbon citrate.
3. During the cycle, two carbons are oxidized to CO_2, regenerating the four-carbon oxaloacetate needed to continue the cycle.
4. Four reactions in the cycle transfer a pair of electrons to NAD^+ to form NADH, or to FAD^+ to form $FADH_2$.
5. Reaction intermediates in the TCA cycle are common compounds generated in other catabolic reactions making the TCA cycle the central metabolic pathway of the cell.
  C. The importance of reduced coenzymes in the formation of ATP:
1. The reduced coenzymes $FADH_2$ and NADH are the primary products of the TCA cycle.
2. NADH formed during glycolysis enters the mitochondria via malate-aspartate or glycerol phosphate shuttles.
3. Electrons pass from $FADH_2$ or NADH to O_2, the terminal electron acceptor, through a chain of carriers in the inner membrane.
4. As electrons move through the electron-transport chain, H^+ are pumped out across the inner membrane.
5. ATP is formed by the controlled movement of H^+ back across the membrane through an ATP-synthesizing enzyme.
6. The coupling of H^+ translocation to ATP synthesis is called chemiosmosis.

7. Three molecules of ATP are formed from each pair of electrons donated by NADH. Two molecules of ATP are formed for each pair of electrons donated by $FADH_2$.

D. The Human Perspective: The role of anaerobic and aerobic metabolism in exercise

1. ATP hydrolysis increases 100-fold during exercise, rapidly exhausting the ATP "on hand."
2. Muscles use stored creatine phosphate (CrP) to rapidly generate ATP but must rely on aerobic or anaerobic synthesis of new ATP for sustained activity.
3. Fast-twitch muscle fibers contract rapidly, have few mitochondria and produce ATP anaerobically.
   a. Anaerobic metabolism produces fewer ATPs per glucose but produces them very fast.
   b. Anaerobic metabolism rapidly depletes available glucose and builds up lactic acid which reduces cellular pH.
4. Slow-twitch fibers contract slowly, have many mitochondria and produce most of their ATP by aerobic metabolism.
   a. Aerobic metabolism initially uses glucose as a substrate.
   b. Free fatty acids are oxidized during prolonged exercise.
5. The ratio of fast- to slow-twitch fibers is variable and depends on the normal function of the muscle.

IV. The Role of Mitochondria in the Formation of ATP

A. ATP can be formed by substrate level phosphorylation or oxidative phosphorylation.

B. Oxidation-reduction potentials:

1. Strong oxidizing agents have a high affinity for electrons; strong reducing agents have a weak affinity for electrons.
2. Redox reactions $A_{(ox)} + B_{(red)}$ $A_{(red)} + B_{(ox)}$ are accompanied by a decrease in free energy: $\Delta G^{\circ\prime} = -nF \Delta E'_o$.
3. The transfer of electrons causes charge separation that can be measured as an oxidation-reduction (redox) potential.
4. The voltage change ($\Delta E'_o$) for the oxidation of NADH to NAD^+ by O_2 is +1.14 V, equal to a $\Delta G^{\circ\prime}$ of -52.6 kcal/mol, enough to synthesize several ATPs ($\Delta G^{\circ\prime}$ = +7.3 kcal/mol).
5. Oxidation of isocitrate, α-ketoglutarate and malate are coupled to the reduction of NAD^+. Oxidation of succinate is coupled to the reduction of FAD.

C. Electron transport:

1. Electrons move through the inner membrane via a series of carriers of decreasing redox potential.
2. Types of electron carriers:
   a. Flavoproteins are polypeptides bound to either FAD or FMN.
   b. Cytochromes contain heme groups bearing Fe or Cu metal ions.
   c. Ubiquinone (coenzyme Q) is a lipid-soluble molecule made of five-carbon isoprenoid units.
   d. Iron-sulfur proteins contain Fe in association with inorganic sulfur rather than heme.
3. Electron-transport complexes:
   a. Complex I (NADH dehydrogenase) catalyzes transfer of electrons from NADH to ubiquinone and transports four H^+ per pair.
   b. Complex II (succinate dehydrogenase) catalyzes transfer of electrons from succinate to FAD to ubiquinone without transport of H^+.

c Complex III (cytochrome bc_1) catalyzes the transfer of electrons from ubiquinone to cytochrome c and transports four H^+ per pair.
d. Complex IV (cytochrome c oxidase) catalyzes transfer of electrons to O_2 and transports H^+ across the inner membrane.
(1) Cytochrome c oxidase is a large well studied complex that adds four electrons to O_2 to form two molecules of H_2O.
(2) The metabolic poisons CO, N_3^- and CN^- bind catalytic sites in Complex IV.
4. Electrons donated by NADH enter through complex I. When $FADH_2$ is the donor, electrons enter through complex II.
5. Free energy released at sites in complexes I, III and IV is sufficient to move protons across the inner membrane.
6. Ubiquinone and cytochrome c are independent of the four complexes.

V. Translocation of Protons and the Establishment of a Protonmotive Force
A. There are two components to the proton gradient:
1. The concentration difference between the matrix and intermembrane space creates a pH gradient (ΔpH).
2. The voltage difference resulting from the separation of charge across the membrane creates an electric potential gradient (Ψ).
3. Approximately 80% of the protonmotive force ($\Psi + \Delta$pH) is due to voltage, 20% to the proton gradient.
B. Dinitrophenol (DNP) uncouples glucose oxidation and ATP formation by increasing the permeability of the inner membrane to H^+, dissipating the proton gradient.
C. Differences in uncoupling protein (UCP-2) among individuals may account for differences in metabolic rate.

VI. The Machinery for ATP Formation
A. Coupling factor 1, the spheres on the matrix side of the inner membrane, is a reversible ATPase (ATP synthase).
B. The structure of the ATP synthase:
1. The F_1 particle is the catalytic subunit represented by the beads seen on the inner membrane in electron micrographs.
2. The F_0 particle attaches to F_1 and is embedded in the inner membrane.
C. The basis of ATP formation according to the binding change mechanism:
1. In the "binding-change mechanism" hypothesis, movement of protons through the ATP synthase alters the binding affinity of the active site for ADP, P_i and ATP.
a. Energy is required to bind the substrates to the enzyme.
b. In the bound state, the addition of P_i to ADP occurs spontaneously.
2. Binding sites on the catalytic subunit can be tight, loose, or open.
3. ATP is synthesized through rotational catalysis where the stalk of ATP synthase rotates relative to the head.
4. There is much structural and experimental evidence to support the binding change mechanism.
5. Using the proton gradient to drive the catalytic machinery:
a. The c subunits of the F_0 portion of the ATPase form a ring.
b. Protons moving through the membrane rotate the c ring.
c. Torque of the rotating ring drives the synthesis of ATP.
D. Other roles for the protonmotive force in addition to ATP synthesis:
1. The H^+ gradient drives transport of ADP into and ATP out of the mitochondria.

2. Transport of Ca^{2+} and some proteins into the mitochondria is energized by the protonmotive force.
3. ADP level is the most important factor controlling the respiration rate.

VII. Peroxisomes
A. Peroxisomes and glyoxysomes are microbodies that perform oxidative metabolism and import proteins post translationally from the cytoplasm.
B. Peroxisomes oxidize a number of organic compounds and synthesize cholesterol, bile acids and plasmalogens.
C. Hydrogen peroxide (H_2O_2) formed in peroxisomes is a reactive and toxic compound broken down by the enzyme catalase.
D. Peroxisomes form by splitting from preexisting organelles, they import preformed proteins, and they engage in oxidative metabolism.
E. Glyoxysomes convert stored fatty acids to glucose for use by newly developing plants.

VIII. The Human Perspective: Diseases That Result from Abnormal Mitochondrial or Peroxisomal Function
A. Mitochondria
   1. A variety of disorders are known that are caused by abnormalities in mitochondria structure and function.
B. Peroxisomes
   1. Patients with Zellweger syndrome lack peroxisomal enzymes due to defects in translocation of proteins from the cytoplasm into the peroxisome.
      a. Defects include an inability to metabolize substrates, reduced ATP formation, and failure to stop oxidizing substrate when ADP levels are low.
      b. Muscle and nerve tissues are most severely affected due to their high metabolic rates.
   2. Adrenoleukodystrophy is caused by the lack of a single peroxisomal enzyme, allowing fatty acids to accumulate in the brain and damage the myelin sheaths of nerves.

IX. Experimental Pathways on the Web: Coupling of Oxidation to Phosphorylation
A. Mitchell, in 1961, first proposed the chemiosmotic theory to explain the coupling of oxidation and phosphorylation.
   1. The pH and electrical gradient resulting from transport of protons links oxidation to phosphorylation.
   2. Carriers accepting both H^+ and electrons should be located on the inner side of the membrane to take up protons.
   3. When electrons are passed to carriers only able to accept electrons, the H^+ is translocated across the inner membrane.
B. A number of testable predictions followed from Mitchell's chemiosmotic hypothesis.
   1. Addition of O_2 to oxygen-starved mitochondria led to the predicted release of H^+ in the medium.
   2. The contribution of pH and electric potential to the protonmotive force was measured.
   3. The number and mechanism of H^+ translocations at each site have been vigorously debated and are still not completely understood.
   4. The generation of artificial proton gradients in chloroplasts proved that a pH gradient alone was sufficient to power phosphorylation.
   5. Reconstitution of bacteriorhodopsin (a light-driven H^+ pump) and ATP synthase into vesicles produced ATP upon illumination.
   6. These last experiments proved that electrons transport and ATP synthesis are not directly coupled.

Key Figure

Figure 5.30. Summary of the major activities that occur during aerobic respiration in a mitochondrion.

Questions for Thought:

1. Can you add a feature to this diagram that illustrates the presence of a protonmotive force showing both its voltage and pH gradient ($\Psi + \Delta pH$) components? Be sure to indicate the direction of both components relative to the membrane.
2. What is the source of the four substances shown entering the mitochondrion in this diagram, namely, P_i, H^+, pyruvic acid, and ADP^{3-}?
3. Which of these components is inhibited by carbon monoxide? What about actinomycin A?
4. The diagram shows one each of the four major complexes of the electron-transport chain and the ATP synthase, and four molecules of cytochrome c. Is there any reason why the ratios of these proteins and protein complexes must be as shown here? Why or why not?

Review Problems

Short Answer

1. Give the significance (not the definition) of the following terms or phrases. Say what they do or why they are important. For example:
   protonmotive force: *provides the energy for the synthesis of ATP.*
   a. evolutionary success of the cyanobacteria

   b. cristae

   c. porins

   d. spatial arrangement of carriers in the electron-transport chain

   e. only 20% of Δp is ΔpH

   f. inner mitochondrial membrane is impermeable to H^+

   g. levels of ADP in the mitochondria

   h. fermentation

   i. catalase in the peroxisomes

   j. Na^+, K^+ ATPase from red blood cells can be made to synthesize ATP

2. Compare and contrast the following:

   a. inner vs. outer mitochondrial membranes

   b. "high energy" vs. "low energy" electrons

   c. substrate level vs. oxidative phosphorylation

   d. chemical vs. electrical gradient

   e. redox loop vs. proton pump

Multiple Choice

1. The fact that mitochondria were known in the nineteenth century to be osmotically active was evidence that:
   a. the mitochondrial matrix is hypotonic to the cellular cytoplasm.
   b. the mitochondrial matrix is hypertonic to the cellular cytoplasm.
   c. mitochondria were derived from a symbiotic bacterium.
   d. mitochondria are surrounded by a semipermeable membrane.

2. Which cell type is likely to have the most mitochondria?
   a. fast-twitch muscle
   b. slow-twitch muscle
   c. liver cells
   d. adipose cells

3. The composition of the inner mitochondrial membrane is most like that of:
   a. the outer mitochondrial membrane.
   b. the cell's plasma membrane.
   c. some bacterial plasma membranes.
   d. the nuclear membrane.

4. Which of the following statements about the outer mitochondrial membrane is not true?
   a. The outer membrane is impermeable to hydrogen ions.
   b. The outer membrane is highly permeable to substances of small molecular weight.
   c. The outer membrane is about 50% lipid and 50% protein by weight.
   d. The outer membrane contains porins.

5. Which of the following would not be found as part of the mitochondrial matrix?
   a. ribosomes
   b. DNA
   c. enzymes
   d. All of the above can be found in the mitochondrial matrix.

6. In what form does the product of glycolysis enter the TCA cycle?
   a. acetyl CoA
   b. pyruvate
   c. NADH
   d. glucose

7. Why is the TCA cycle the central pathway of the cell?
   a. It occurs in the center of the cell.
   b. Its metabolites are the same compounds made by most catabolic pathways.
   c. All other metabolic pathways depend upon it.
   d. None of the above.

8. Which of the following are reduced coenzymes?
   a. NADH and $FADH_2$
   b. NAD^+ and FAD
   c. ATP and GTP
   d. coenzyme A and ubiquinone

9. Which of the following is not a source of energy in active muscle cells?
   a. creatine phosphate
   b. ATP
   c. lactic acid
   d. glucose

10. Which of the following is *not* a feature of oxidative phosphorylation?
    a. direct transfer of phosphate from a substrate molecule to ADP
    b. an electrochemical gradient across the inner mitochondrial membrane
    c. a membrane-bound ATP synthase
    d. a protonmotive force

11. Given the following redox couples and standard oxidation-reduction potentials:

	E'_0 (V)
2 cytochrome c_{ox} + 2 e⁻ ⇔ 2 cytochrome c_{red}	+0.254 V
2 cytochrome a_{3ox} + 2 e⁻ ⇔ 2 cytochrome a_{3red}	+0.385 V

    which way would electrons move between these couples under standard conditions?
    a. From cytochrome c_{red} to cytochrome a_{3ox}
    b. From cytochrome a_{3red} to cytochrome c_{ox}
    c. From cytochrome c_{red} to cytochrome c_{ox}
    d. From cytochrome a_{3red} to cytochrome a_{3ox}

12. Why do some redox reactions in cells result in the transfer of electrons from a reductant with a higher standard redox potential to an oxidant with a lower standard redox potential?
    a. Redox potentials are defined under standard conditions, and cellular conditions are typically not standard conditions.
    b. The rules governing oxidation and reduction apply only *in vitro*, and not *in vivo*.
    c. Electrons always go from reductants with high redox potentials to oxidants with low redox potentials. It's the rule, not the exception.
    d. None of the above.

13. What is the source of free energy (ΔG) for moving protons out of mitochondria against their concentration and electrical gradients?
   a. glycolysis
   b. creatine phosphate
   c. the redox reactions of electron transport
   d. ATP

14. Which of the following is *not* an electron carrier class in the electron transport chain?
   a. flavoproteins
   b. cytochromes
   c. iron-sulfur proteins
   d. cytochrome c oxidase

15. If complex III were incorporated into an artificial lipid vesicle in order to demonstrate its proton translocating ability in isolation, which of the following would be an appropriate electron donor?
   a. cytochrome a_{red}
   b. cytochrome c_{red}
   c. UQH_2
   d. cytochrome c oxidase

16. In glycolysis, ATP is formed by the transfer of a high energy phosphate from 1,3-bisphosphoglycerate to ADP. No such high energy phosphate donor has ever been isolated in mitochondria. Why not?
   a. The techniques for isolating the phosphate donor are not refined enough .
   b. No such phosphate donor exists.
   c. The high energy phosphate donor is very short-lived and difficult to isolate.
   d. None of the above.

17. Which of the following could act as an "uncoupler" of electron transport and ATP synthesis?
   a. the F_o base piece of the ATP synthase (without the F_1 subunit)
   b. dinitrophenol
   c. neither a nor b
   d. both a and b

18. Which of the following best describes the mechanism by which ATP is synthesized by the ATP synthase using the energy of proton movement?
   a. Protons move through the F_o subunit and thereby allow ADP and P_i to bind the enzyme.
   b. Protons moving through the F_o subunit rotate one subunit relative to the other, enabling the bound ATP to be released..
   c. Protons bind to the F_1 subunit and electrostatically join ADP and P_i.
   d. Protons allow substrate-level phosphorylation to occur, much like that in glycolysis.

19. Using differential centrifugation, you are trying to isolate peroxisomes and glyoxysomes from a mixture of cellular organelles. After a few centrifugation steps, you think you may have a relatively pure suspension. How might you determine that your suspension does, indeed, have these organelles?
   a. Do an assay for the enzyme catalase.
   b. Do an assay for the enzyme succinate dehydrogenase.
   c. Look in the microscope for a double-membrane bound organelle.
   d. Assay your suspension for nucleic acid.

20. Which of the following experimental results does not support Mitchell's chemiosmosis theory?
   a. Electron transport in isolated mitochondria was shown to result in acidification of the medium.
   b. Addition of dinitrophenol to isolated mitochondria was shown to inhibit electron transport, but had no effect on ATP synthesis.
   c. Addition of dinitrophenol to isolated mitochondria during electron transport was shown to inhibit acidification of the medium.
   d. An artificial proton gradient across the inner mitochondrial membrane drives ATP synthesis in the absence of electron transport.

Problems and Essays

1. List three mitochondrial features consistent with the view that mitochondria first arose as bacterial symbionts residing within the cytoplasm of other prokaryotic host cells.

2. a. Write the overall, simplified, balanced reaction for the TCA cycle.

   b. In the first reaction of the TCA cycle, oxaloacetate combines with acetyl CoA to form citrate, yet there is no net disappearance of oxaloacetate. Explain.

Use this table to answer questions (3) through (5):

Oxidants		Reductants	E_o' (V)
$NAD^+ + 2H^+ + 2e^-$	$\Leftrightarrow$	$NADH + H^+$	-0.320
2 cytochrome $b_{ox} + 2e$	$\Leftrightarrow$	2 cytochrome b_{red}	+0.030
ubiquinone$_{ox}$ + $2H^+ + 2e^-$	$\Leftrightarrow$	ubiquinone$_{red}$	+0.100
2 cytochrome $c_{ox} + 2e^-$	$\Leftrightarrow$	2 cytochrome c_{red}	+0.254
$1/2\ O_2 + 2H^+\ 2e^-$	$\Leftrightarrow$	H_2O	+0.816

3. a. Write the simplified, balanced reaction that would occur between ubiquinone and cytochrome c under standard conditions.

    b. Calculate the standard free-energy change, $\Delta G^{\circ\prime}$, of the reaction in part (a).

    c. Would this reaction yield enough free energy to synthesize a mole of ATP under standard conditions?

4. a. In the electron-transport chain, electrons move between ubiquinone and cytochrome b. Which molecule would act as the oxidant and which would be the reductant under standard conditions?

    b. Could the roles you've assigned in part (a) be reversed under cellular conditions? If so, what conditions would have to be maintained?

5. a. Calculate the free energy released when one mole of NADH reduces one mole of oxygen under standard conditions.

    b. This reaction occurs as part of oxidative metabolism, albeit quite indirectly. (Recall from Chapter 3 that the free-energy change of a reaction is independent of the pathway the reaction takes.) Based on your answer to part (a), propose an explanation for why cells have evolved the electron-transport chain to carry electrons from NADH to oxygen rather than a direct redox reaction between the pair.

6. Mitchell and Moyle determined that the total protonmotive force across the inner mitochondrial membrane was 230 mV, of which 80% was due to the electrical potential. What was the pH gradient across the membrane?

7. Based on the chemiosmotic mechanism of oxidative phosphorylation, briefly predict the results of the following experiments:

   a. When inside out submitochondrial particles are engaged in electron transport, the interior of the vesicles_____.

   b. When electron transport is blocked in intact mitochondria, and an artificial pH gradient is imposed by lowering the pH of the medium, the synthesis of ATP_____.

   c. FCCP is a drug that partitions itself into membranes and opens a channel specific for the passage of protons. When FCCP is added to a suspension of mitochondria, electron transport_____, and ATP synthesis_____.

   d. Under anaerobic conditions, the pH gradient_____(decreases, increases or stays the same) relative to aerobic conditions.

Web Explorations

Everyone is familiar with white adipose tissue, also called fat tissue or white fat. White fat stores chemical energy in the form of triacylglycerols and may provide insulation to warm blooded animals. But some mammals also have a second kind of adipose tissue, called brown adipose tissue (BAT). BAT contains a natural uncoupling protein, called thermogenin, in the inner mitochondrial membrane. Mitochondria with thermogenin oxidize foodstuffs, but the energy released is converted to heat instead of ATP.

Using the World Wide Web, find out how thermogenin uncouples electron transport and ATP synthesis. What kinds of mammals have brown fat as newborns? What kinds of mammals have brown fat as adults? In what ways do the effects of thermogenin in BAT mimic the symptoms of the patient deescribed by Rolf Luft (The Human Perspective, page 212)?

Notes:

6 Photosynthesis and the Chloroplast

Learning Objectives

When you have finished this chapter, you should be able to:
1. List the major evolutionary steps leading from the earliest autotrophs to higher plants.
2. Describe the chloroplast, paying particular attention to membranes and how chloroplast membranes function in photosynthesis.
3. Distinguish between the light-dependent reactions and the light-independent reactions of photosynthesis. Summarize the major steps in each of these pathways, including their substrates and ultimate products.
4. List the major electron carriers in photosynthesis. Describe how the oxidation and reduction energy of these carriers is stored, and then used to make ATP, NADPH, and ultimately carbohydrate.
5. Distinguish between productive and nonproductive reactions that occur in the chloroplast, and describe some of the adaptations that have evolved to maximize the productive pathways and minimize the nonproductive ones.

Key Terms and Phrases

heterotrophs
photoautotrophs
thylakoid
stroma
light-independent reactions
excited state
chromophore
accessory pigments
reaction center
photosystem II (PSII)
primary electron acceptor
photolysis
plastocyanin
phylloquinone (A1)
cyclic photophosphorylation
Rubisco
transpiration
bundle sheath cells

autotrophs
photosynthesis
grana
stroma thylakoids (stroma lamellae)
photon
chlorophyll
absorption spectrum
carotenoids
antenna
P680
light-harvesting complex II (LHCII)
pheophytin
appressed
noncyclic photophosphorylation
C_3 pathway
ribulose bisphosphate carboxylase
stomata
CAM plants

chemoautotrophs
chloroplast
lumen
light-dependent reactions
ground state
pigment
action spectrum
photosynthetic unit
photosystem I (PSI)
P700
Z scheme
plastoquinone
nonappressed
CO_2 fixation
C_4 (Hatch-Slack) pathway
photorespiration
mesophyll cells
photoinhibition

Reviewing the Chapter

I. Introduction

    A. The earliest living organisms were heterotrophs that obtained organic substrates from the environment.

    B. Autotrophs manufacture organic nutrients from CO_2, and either H_2S or H_2O.

    C. Synthesis of complex molecules from CO_2 requires energy.

        1. Chemoautotrophs use energy stored in inorganic molecules to synthesize organic compounds.

        2. Photoautotrophs use radiant energy to carry out photosynthesis, the transformation of sunlight into chemical energy.

    D. Photosynthesis converts low energy electrons from a donor molecule into high energy electrons that can be used in anabolic reactions.

        1. The first photoautotrophs used hydrogen sulfide (H_2S) as an electron source.

        2. Water (H_2O) is a more abundant source of electrons; photosynthesis using water produces molecular oxygen (O_2) as a by-product.

II. Chloroplast Structure and Function

    A. In eukaryotic cells, photosynthesis takes place in large, membranous cytoplasmic organelles called chloroplasts.

    B. Chloroplasts are thought to be ancient cyanobacteria that took up residence inside a nonphotosynthetic proeukaryotic cell.

    C. Chloroplasts have a double membrane, like the mitochondria.

        1. The outer membrane contains porin and is permeable to large molecules.

        2. The inner membrane contains light-absorbing pigment, electron carriers and ATP-synthesizing enzymes.

    D. The inner membrane is folded into flattened sacs called thylakoids, arranged in stacks called grana.

    E. Chloroplasts are semiautonomous, self-replicating organelles containing their own small, circular DNA molecules.

    F. Thylakoid membranes contain a high percentage of unsaturated neutral glycolipids which increase fluidity and facilitate protein mobility.

III. An Overview of Photosynthetic Metabolism

    A. Photosynthesis is an oxidation-reduction reaction transferring an electron from water to carbon dioxide: $6\,CO_2 + 12\,H_2O \longrightarrow C_6H_{12}O_6 + 6\,H_2O + 6\,O_2$

    B. Experiments using ^{18}O demonstrated that the O_2 released during photosynthesis came from two molecules of H_2O, not CO_2, as previously believed.

    C. During photosynthesis, water is oxidized to form oxygen; in respiration, oxygen is reduced to form water.

        1. Respiration removes high energy electrons from reduced organic substrates to form ATP and NADH.

        2. Photosynthesis boosts low energy electrons to form ATP and NADPH, which are then used to reduce CO_2 to carbohydrate.

    D. Photosynthesis occurs in two stages:

        1. Light-dependent (light) reactions absorb light, converting it into chemical energy in the forms of ATP and NADPH.

        2. Light-independent (dark) reactions use the energy of ATP to form carbohydrate.

IV. The Absorption of Light

    A. Absorption of photons lifts electrons from inner to outer orbitals, elevating molecules from the ground state to an excited state.

        1. The energy in a mole of photons depends on the wavelength of the light.

2. The energy required to shift electrons varies for different molecules.
3. Therefore, molecules absorb specific wavelengths of light.
B. Energy from excited molecules can be re-radiated at a lower wavelength (fluorescence) or, in the case of chloroplasts, transferred to electron carriers.
C. Photosynthetic pigments are molecules with chromophores, groups capable of absorbing light.
1. Chlorophyll contains a porphyrin ring that absorbs light and a hydrophobic tail anchoring it to the chloroplast membranes.
2. A conjugated system of single and double bonds in the porphyrin ring creates a cloud of delocalized electrons that strongly absorb light and broaden the absorption spectrum.
D. Different photosynthetic organisms have different classes of chlorophyll.
1. Chlorophyll a is present in all oxygen-producing photosynthetic organisms.
2. Chlorophyll b is present in higher plants and green algae.
3. Chlorophyll c is found in brown algae, diatoms and photosynthetic protozoa.
E. Accessory pigments broaden the range of absorbed light able to stimulate photosynthesis and protect photosynthetic molecules from reactive forms of oxygen.

V. Photosynthetic Units and Reaction Centers
A. Each photosynthetic unit contains 300 chlorophyll molecules that harvest light.
B. Polypeptides facilitate energy transfer by holding pigment molecules in fixed orientations.
C. Excitation energy is transferred to a reaction-center chlorophyll that passes electrons to electron acceptors.
D. Oxygen formation: Coordinating the action of two different photosynthetic systems
1. Two photosystems act in series to boost electrons from H_2O to NADPH.
   a. Photosystem II (PSII) boosts electrons from below the energy level of water to a point midway between water and $NADP^+$.
   b. Photosystem I (PSI) boosts electrons to a level above $NADP^+$.
2. The reaction center of PSII is referred to as P680, and that of PSI as P700.
3. The Z scheme describes the flow of electrons from water to $NADP^+$.
E. PSII operations: Obtaining electrons by splitting water:
1. PSII uses the absorbed light energy to remove electrons from water and to generate a proton gradient.
2. Two peptides, D1 and D2, bind the P680 chlorophyll molecule and perform reactions required to oxidize water.
3. Light is harvested by a pigment-protein complex called the light-harvesting complex II (LHCII).
4. The flow of electrons from PSII to plastoquinone (PQ):
   a. Harvested energy is passed through a core antenna complex to P680.
   b. Transfer of energy from P680 to a primary electron acceptor (Pheo) generates a pair of opposite charged species $P680^+$ and $Pheo^-$.
   c. $P680^+$ and $Pheo^-$ are moved to opposite sides of the thylakoid membrane.
   d. $Pheo^-$ passes an electron to PQ, which passes it to another PQ molecule.
   e. The electron is being moved to the stromal side of the membrane.
5. The flow of electrons from water to PSII:
   a. $P680^+$ is a powerful oxidant capable of pulling electrons from water (photolysis).
   b. The four electrons required to form 1 molecule of O_2 are transferred in successive cycles through $P680^+$ to four Mn^{2+} ions, allowing the oxygen-evolving complex of PSII to catalyze the removal of four electrons from two molecules of H_2O.
   c. Protons produced in photolysis remain in the lumen of the thylakoids.

                  d. Oxygen produced as a waste product of photolysis is released into the environment.

   F. From PSII to PSI
      1. Reduced PQH_2 (2 for every molecule of O_2 produced) binds cytochrome b_6f and releases protons into the lumen of the thylakoid.

      3. Plastocyanin transfers electrons to $P700^+$, the reaction center of PSI.
   G. PSI operations: The production of NADPH:
      1. The PSI core contains 12 different polypeptides which bind 96 chlorophylls, 22 carotenoids and 9 small cofactors.
      2. The photochemical events that take place in PSI begin with the absorption of photons by the chlorophyll and ß-carotene molecules in the core complex.
      3. Photons harvested by antenna pigments in PSI (LHCI) oxidize chlorophyll a in P700, forming $P700^+$.
      4. The electrons formed by oxidation of P700 pass to another molecule of chlorophyll a acting as a primary electron acceptor (A_0^-).
      5. The redox potential of the $P700^+/A_0^-$ pair reduces $NADP^+$.
      6. Transfer of electrons from $P700^{+\prime}$ on the luminal side, through intermediates to $NADP^+$, on the stromal side, adds to the proton gradient.
      7. Electrons from PSI can be diverted to reduce nitrate, ammonia or sulfate forming compounds necessary for life.
   H. An overview of photosynthetic electron transport:
      1. For each 8 photons absorbed, 2 moles H_2O are split, 2 moles NADPH are produced, and one mole O_2 is produced.
      2. Electron transport also produces a proton gradient across the thylakoid membrane.
   I. Killing weeds by inhibiting electron transport:
      1. Certain herbicides reduce O_2 to toxic oxygen radicals that kill plants.
      2. Some herbicide residues are toxic to humans.

VI. Photophosphorylation
   A. The ATP synthase in chloroplasts is constructed of polypeptides homologous to mitochondrial enzymes and believed to function by a similar mechanism.
   B. The synthase headpieces (CF_1) are aggregated in exposed regions of the grana stacks and project into the stroma so H^+ move through the synthase down the H^+ concentration gradient from lumen to stroma.
   C. Isolated chloroplasts engaged in photosynthesis alkalinize the incubation medium creating a pH gradient of 3 to 4 pH units.
   D. The electromotive force created in chloroplasts by electron transport does not have a large voltage component because other ions are transported simultaneously.
   E. The movement of electrons during photosynthesis, a process resulting in the formation of oxygen, is linear and called noncyclic photophosphorylation.
   F. Noncyclic versus cyclic photophosphorylation:
      1. Cyclic photophosphorylation is carried out independently by PSI and involves the recycling of the electrons from ferredoxin back to the electron-deficient reaction center.
      2. The degree of cyclic vs. non-cyclic photophosphorylation depends on the energy needs of the cell at the time.

VII. Carbon Dioxide Fixation and the Synthesis of Carbohydrate
   A. The movement of carbon in the cell can be followed during photosynthesis using $[^{14}C]O_2$ as a tracer.

B. Carbohydrate synthesis in C_3 plants:
   1. Carbon is rapidly incorporated into carbohydrate.
   2. The first compound to be identified was 3-phosphoglycerate (PGA), a three-carbon intermediate in glycolysis. The pathway was called the C_3 pathway.
   3. CO_2 is condensed with a five-carbon compound, ribulose 1,5-bisphosphate (RuBP) to form a six-carbon compound that splits into two molecules of PGA.
   4. The condensation and splitting are two activities of a single enzyme, ribulose bisphosphate carboxylase (Rubisco).
C. The C_3 pathway (called the Calvin cycle) includes three basic steps:
   1. Carboxylation of RuBP and splitting to form PGA.
   2. Reduction of PGA to glyceraldehyde 3-phosphate (GAP) using NADPH and ATP formed by electron transport.
   3. Regeneration of RuBP.
D. The GAP molecules can be used in the cytosol as metabolic substrates, or converted to sucrose and carried to nonphotosynthetic tissues.
E. Alternatively, GAP can be converted to starch in the chloroplast where it is stored for use during periods when photosynthesis has stopped.
F. The formation of carbohydrate is expensive (12 NADPH and 18 ATP per six-carbon sugar) reflecting the fact that CO_2 is the most highly oxidized form of carbon.
G. Enzymes of the Calvin cycle can be found in other pathways.
H. Redox control of the Calvin cycle is light dependent:
   1. Key enzymes of the Calvin cycle are only active when ATP and NADPH are being produced by photosynthesis.
   2. Some electrons used to reduce $NADP^+$ in light dependent PSI operations are transferred to thioredoxin
   3. Thioredoxin reduces disulfide bridges (-S-S-) into sulfhydryl groups (-SH) in select Calvin cycle enzymes.
   4. In the dark reduction of thioredoxin ceases and calvin cycle enzymes revert to an oxidized (-S-S-) inactive state.
I. Photorespiration:
   1. During carbon fixation, the active site on Rubisco binds RuBP, making RuBP susceptible to attack by either CO_2 or O_2.
   2. Photorespiration is the utilization of O_2 to form glycolate and the subsequent release of CO_2 in the mitochondria.
   3. Whether photorespiration or photosynthesis occurs depends on the CO_2/O_2 ratio.
J. Peroxisomes and Photorespiration:
   1. One product of photorespiration, glycolate, is shunted from the chloroplast into perixisomes.
   2. Peroxisomes convert glycolate into amino acids with the loss of one CO_2.
K. Carbohydrate synthesis in C_4 plants:
   1. C_4 plants combine CO_2 with the three-carbon compound phosphoenolpyruvate (PEP) to form four-carbon skeletons, oxaloacetate or malate.
   2. The first step in the C_4 (or Hatch-Slack) pathway is the attachment of CO_2 to PEP by phosphoenolpyruvate carboxylase.
   3. The C_4 plants are primarily tropical grasses and can utilize CO_2 at levels far below the 50 ppm threshold where photosynthesis ceases in C_3 plants.
   4. In hot, dry environments C_4 plants can acquire enough CO_2 for photosynthesis while keeping their stomata partially closed to prevent excessive water loss.

5. C_4 products are transported to bundle sheath cells, where they are protected from the atmosphere. The CO_2 is cleaved off and concentrations of CO_2 high enough for utilization by Rubisco are attained.

L. Carbohydrate synthesis in CAM plants:
1. CAM (Crassulacean acid metabolism) plants fix CO_2 at night using PEP carboxylase.
2. Stomata remain closed during daylight hours to conserve water.

Key Figure

Figure 6.5. An overview of the energetics of photosynthesis and aerobic respiration.

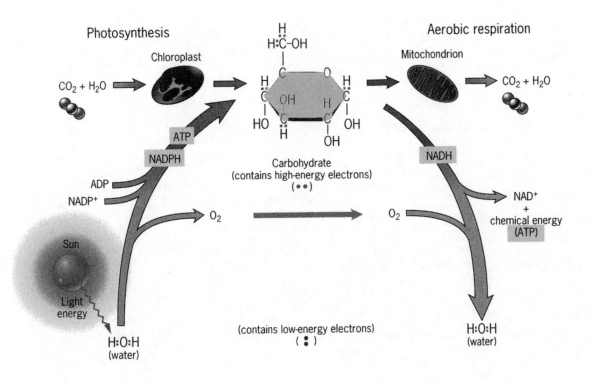

Questions for Thought:

1. The textbook describes this figure as an overview of the thermodynamics of photosynthesis and aerobic respiration. If you imposed x-y axes on this figure, which axis would represent energy? How would you label the other axis?
2. Describe the experiments that proved that the molecular oxygen from photosynthesis is derived from water and not carbon dioxide.
3. Can you list all of the compounds shown in this figure in order of increasing redox potential?
4. While the van Niel experiment that you described in question 2 implied that aerobic respiration is the reverse of photosynthesis, there are many ways in which the two pathways differ. Describe as many differences between these to pathways as you can.

Review Problems

1. Give the significance (not the definition) of the following terms or phrases. Say what they do or why they are important. For example:
   photoautotrophs: *capture light energy from the sun. The rest of the biotic world depend upon them for energy.*

   a. the switch from H_2S to H_2O as a source of electrons for photosynthesis

   b. antenna

   c. fluidity of the thylakoid membranes

   d. carotenoids

   e. two photosystems, PS I and PS II

   f. alternate routes that electrons can take from ferredoxin (other than to NADPH)

   g. thylakoid membranes are impermeable to protons

   h. Rubisco binds to RuBP, not CO_2 or O_2

   i. peroxisomes are closely associated with chloroplasts

   j. C_4 and CAM pathways

2. Compare and contrast the following:

    a. chemoautotrophs vs. photoautotrophs

    b. grana thylakoids vs. stroma thylakoids

    c. light-dependent vs. light-independent reactions

    d. cyclic vs. noncyclic photophosphorylation

    e. mesophyll cells vs. bundle sheath cells

Multiple Choice

1. Of the following, the most decisive event in biological evolution was:
   a. the evolution of heterotrophy.
   b. the evolution of chemoautotrophy.
   c. the evolution of photoautotrophy.
   d. the evolution of carbon dioxide.

2. The products of the light reactions of photosynthesis are:
   a. ATP and NAD^+.
   b. ATP and NADPH.
   c. glucose.
   d. sucrose.

3. The energy content of a photon of light depends upon:
   a. wavelength.
   b. temperature.
   c. the excited state.
   d. whether or not it is emitted in a vacuum.

4. The energy of an excited electron of a chlorophyll molecule:
   a. is released as heat.
   b. is released as fluorescence.
   c. is passed on to an electron acceptor.
   d. All of the above may be true, depending on circumstances.

5. If the reaction-center chlorophyll absorbs light of wavelength 700 nm, then the antenna pigment molecules of that photosynthetic unit must absorb:
   a. light with wavelengths less than 700 nm.
   b. light with wavelengths greater than 700 nm.
   c. light with a wavelength of exactly 700 nm.
   d. light of all wavelengths.

6. In photosystem II, P680 is oxidized to $P680^+$ and pheophytin is reduced to $Pheo^-$. What prevents $Pheo^-$ from reducing $P680^+$ back to P680?
   a. They are not a compatible redox pair.
   b. The opposite charges are moved to different sides of the thylakoid membrane.
   c. The redox potentials of $P680^+$ and $Pheo^-$ make it thermodynamically unfavorable.
   d. None of the above.

7. The driving force for photolysis is:
   a. an electron gradient across the thylakoid membrane.
   b. ATP.
   c. the redox potential of $P680^+$.
   d. NADPH.

8. Which of the following must be true of the chloroplast membranes?
   a. Thylakoid membranes are impermeable to protons.
   b. The ATP synthase is located in the outer chloroplast membrane.
   c. The ATP synthase is located in the inner chloroplast envelope.
   d. All of the above.

9. Plastoquinone is to ubiquinone as the thylakoid membrane is to:
   a. the inner mitochondrial membrane.
   b. the outer mitochondrial membrane.
   c. the stromal membranes.
   d. the inner chloroplast membrane.

10. Which of the following does not directly contribute to the proton gradient across the thylakoid membrane?
    a. photolysis
    b. ferredoxin-$NADP^+$ reductase
    c. cytochrome b_6f complex
    d. plastocyanin

11. Which of the following is not part of cyclic photophosphorylation?
    a. PSII
    b. PSI
    c. ATP synthase
    d. cytochrome b_6f complex

12. Which of the following is not a characteristic of the enzyme ribulose bisphosphate carboxylase?
    a. It links CO_2 to a two-carbon acceptor, resulting in the synthesis of 3-phosphoglyceric acid.
    b. It is the most abundant protein on Earth.
    c. It catalyzes both carbon fixation and photorespiration.
    d. It is soluble in the stroma of the chloroplast.

13. The functional role of glucose in animals is to sucrose in plants as glycogen in animals is to:
    a. glucose in plants.
    b. ATP in plants.
    c. starch in plants.
    d. plants.

14. The most highly oxidized form of carbon is:
    a. carbohydrate.
    b. carbon in protein.
    c. carbon dioxide.
    d. hydrocarbon.

15. What are the most likely products of the reaction catalyzed by Rubisco under normal atmospheric conditions, where O_2 is abundant and CO_2 is low?
    a. 3-phosphoglycerate
    b. RuBP
    c. glycine and serine
    d. 2-phosphoglycolate

16. What products of photorespiration can be salvaged even though it is generally inefficient to do so?
    a. NADPH
    b. Amino acids
    c. ATP
    d. Important components of cell membranes

17. If you wanted to isolate the enzyme phosphoenolpyruvate carboxylase, where would you look?
    a. bundle sheath cells of C_4 plants
    b. mesophyll cells of C_4 plants
    c. chloroplasts of C_3 plants
    d. mitochondria of CAM plants

18. If you wanted to isolate large quantities of malate from CAM plants, when would you look?
    a. early morning
    b. midday on a sunny day
    c. midnight
    d. time of day wouldn't matter

19. Why might ATP synthase molecules be excluded from appressed membranes?
    a. They need to be near PSI complexes.
    b. They require large amounts of MGDG to function.
    c. They have large CF_1 subunits that interfere with membrane-membrane contact.
    d. They are not excluded from appressed membranes.

20. Of the following observations, which does *not* support the view that appressed and nonappressed photosynthetic membranes differ in protein composition?
    a. Antibodies for PSI bind at different sites than do antibodies for PSII.
    b. The proton gradient across the thylakoid membrane is uniform at all positions within the thylakoid.
    c. The thylakoid membrane is highly fluid, promoting lateral movement of plastiquinone.
    d. Plastoquinone is found both in appressed and nonappressed membranes.

Problems and Essays

1. Given the following oxidation potentials:

Oxidants		Reductants	E_0' (V)
$NADP^+ + H^+ + 2e^-$	$\Leftrightarrow$	NADPH	-0.320
$S° + 2H^+ + 2e^-$	$\Leftrightarrow$	H_2S	+0.030
$1/2\,O_2 + 2H^+ + 2e^-$	$\Leftrightarrow$	H_2O	+0.816

a. How much energy is required for H_2O to reduce one mole of $NADP^+$ under standard conditions? With H_2S as the electron donor?

b. Given the following equations and constants:

$$E = hv \quad and \quad v = \frac{c}{\lambda}$$

$$h = 1.58 \times 10^{-34} \; cal \cdot sec \; \text{(Planck's constant)}$$

$$c = 3 \times 10^8 \; m/sec \; \text{(the speed of light in a vacuum)}$$

$$1 \; mole = 6.02 \times 10^{23}$$

$$1 \; m = 10^9 \; nm$$

How much energy does a mole of photons have at a wavelength of 680 nm? At a wavelength of 485 nm?

c. Assuming that the reactions are taking place under standard conditions and that all of the energy of a mole of photons is captured, can one mole of photons at 680 nm provide enough energy for H_2O to reduce a mole of $NADP^+$? What about a mole of photons at 485 nm?

d. Making the same assumptions as in part (c), how many moles of photons of wavelength 680 nm are required to provide the energy for H_2O to reduce one mole of $NADP^+$?

2. Refer to the Calvin cycle illustrated in figure 6.19. What ratio of ATP to NADPH is required to synthesize one mole of six-carbon sugar? Assuming that noncyclic photosynthesis produces equimolar amounts of ATP and NADPH, how must plants supplement their ATP levels to maintain the appropriate ratio of ATP to NADPH?

3. The proton gradient that drives ATP synthesis is generated at four different points in noncyclic photophosphorylation, as listed below. For each step, indicate whether a gradient is formed by proton translocation across the thylakoid (**PT**), proton consumption on the stromal side of the membrane (**SC**), or proton release on the lumen side (**RL**):

_____photolysis

_____QH_2

_____cytochrome b_6f complex

_____ferredoxin-$NADP^+$ reductase

4. Indicate whether the following experimental results apply to C_3, C_4, and/or CAM plants:

_____a. The plant was incubated in the presence of $^{14}CO_2$, and the first organic compounds in which radioactivity appeared were malate and oxaloacetate.

_____b. The plant was incubated in the presence of $^{14}CO_2$, and the radioactivity appeared both in 3-phosphoglycerate and glycolate.

_____c. The plant was incubated in the presence of $^{14}CO_2$, and newly incorporated radioactivity appeared in organic compounds during the night, but not during the day.

_____d. The leaves of the plant were examined under the microscope and were found to have two concentric cylinders of cells.

5. a. The following are lists of pigments (and their absorption maxima) that form the photosynthetic units of photosystems I and II. For each list, indicate the pigment molecule that is acting as the reaction center. Assuming all the pigments participate, indicate the pathway that energy takes in going to the reaction center.

PSI	PSII
Chl a (660)	Chl a (660)
Chl b (500)	Chl a (670)
Chl a (678)	Chl a (678)
carotenes (490)	Chl b (500)
Chl a (685)	Chl a (680)
Chl a (700)	xanthophylls (450)
Chl a (670)	Chl a (685)
Chl a (690)	

b. Plants with the pigments listed above were used in a series of experiments that measured the effect of wavelength on photosynthesis. When plants were given light of 700 nm, oxygen production ceased, even though the absorption spectrum indicated that some chlorophyll could absorb light of that wavelength. If the 700 nm light was supplemented with a beam of light at 650 nm, oxygen production resumed. Explain.

6. The membranes of the chloroplast play an essential role in photosynthesis. List three ways in which photosynthesis is dependent upon the membranes of the chloroplast.

Web Explorations

One of the side effects of burning fossil fuels in vehicles and industrial settings is an increase in the concentration of carbon dioxide in the atmosphere with a consequential increase in the amount of heat trapped within the Earth's atmosphere. This is called the greenhouse effect. Some have argued, however, that high atmospheric CO_2 could act as a kind of CO_2 fertilizer, increasing the rate of carbon fixation by trees. The Earth's plants sequester a large quantity of carbon dioxide in the form of leaves, wood, roots, and stems, but are they capable of absorbing the excess CO_2 that is threatening to raise global temperatures? In this web exploration you will examine the greenhouse effect and its causes and make a reasoned argument for why or why not primary productivity (i.e., phtotosynthesis) can mitigate increased atmospheric CO_2 and prevent global warming.

Begin by exploring the World Wide Web for information on the greenhouse effect. How does CO_2 contribute to increased global temperature? How have atmospheric CO_2 levels changed in thepast 100 years? Now look for evidence that the photosynthetic rates of plants—forests in particular—are influenced by increased atmospheric CO_2. In your research be sure to differentiated between long-term and short-term effects. Make a reasoned argument for or against environmentalist action aimed at lowering the levels of atmospheric CO_2.

Interactions Between Cells and Their Environment

Learning Objectives

When you have finished this chapter, you should be able to:

1. Describe the extracellular matrix. Give some examples of tissues whose characteristic properties are determined by their respective extracellular matrices.
2. List the major components of the extracellular matrix, and include the functional role of each.
3. Discuss the families of proteins that mediate cell-to-cell and cell-to-substrate adhesion. Explain the roles of some of these proteins in inflammation and cancer.
4. Describe the cell surface specializations that enhance attachment of cells to substrates and to other cells.
    a. Give examples of cells that express each of these types of connections, respectively.
    b. Distinguish between the functional features of each type of cell-to-cell connection.
5. Describe the origin and role of plant cell walls.

Key Terms and Phrases

glycocalyx	extracellular matrix (ECM)	basement membrane
collagen	proteoglycans	glycosaminoglycans
fibronectin	matrix metalloproteinases (MMPs)	laminin
integrin	RGD sequence	focal adhesions
hemidesmosomes	immunoglobulin superfamily (IgSF)	selectins
NCAM and VCAM	cadherins	adherins junctions
protocadherins	catenins	desmosomes
desmogleins	desmocollins	metastasis
tight junctions	transmembrane signaling	connexin
connexon	occludin	claudins
plasmodesmata	cell walls	microfibrils
hemicellulose	pectins	cellulose
cellulose synthase	hemicellulose	extensins
lignins	primary and secondary cell walls	cell plate

Reviewing the Chapter

I. Introduction
    A. Cells interact with extracellular material and other cells to form tissues.
    B. These interactions mediate diverse activities: cell migration, growth, differentiation and tissue organization, during embryogenesis.

II. The Extracellular Space
    A. The glycocalyx (cell coat) is formed by carbohydrates projecting from membrane lipids and proteins.

97

B. The extracellular matrix (ECM) is an organized network beyond the plasma membrane.
   1. A basement membrane (basal lamina) underlies epithelia and surrounds muscle and fat cells.
      a. The basement membrane maintains epithelial cell polarity.
      b. The basement membrane assists in cell migration.
      c. The basement membrane forms a barrier to macromolecules.
   2. The ECM gives connective tissues their identifiable properties.
C. Components of the ECM:
   1. Collagens are a family of glycoproteins found only in the ECM.
      a. Collagen is a trimer containing three polypeptide chains.
      b. The structure of fibrillar collagens provides high tensile strength.
      c. The three polypeptide chains are wound around each other to form a unique, rodlike triple helix.
      d. Most collagens are fibrillar collagens and because of their widespread distribution any abnormalities can cause serious disorders.
      e. Matrix metalloproteinases degrade proteins of the ECM as part of normal, dynamic turnover.
   2. Proteoglycans (PGs) are protein-polysaccharide complexes with a core protein attached to glycosaminoglycan (GAG) chains.
      a. GAGs have repeating disaccharide structure (A-B-A-B-A-B)
      b. Negatively charged GAGs attract small cations and water, forming a porous, hydrated gel.
   3. Fibronectins are linear arrays of modular polypeptides.
      a. Each modular polypeptide contains about 30 Fn modules.
      b. Fn modules are found in other proteins, as well.
      c. Fibronectin has binding sites for other ECM proteins as well as for cell surface receptors.
      d. Fibronectin guides cells that migrate during development; e.g., neural crest cells.
   4. Laminins are extracellular glycoproteins with binding sites for cell surface receptors, proteoglycans, and other laminins.
      a. Laminins guide the migration of several cell types in development, including primordial germ cells.
      b. Laminins are components of basement membranes.
D. The ECM is a dynamic structure, changing in both time and space.
   1. The ECM can be stretched under tension.
   2. The ECM is continually degraded by matrix metalloproteinases (MMPs), and reconstructed.

III. Interaction of Cells with Noncellular Substrates

A. Integrins are membrane-spanning heterodimers containing two noncovalently linked chains, called α and ß.
   1. Integrins have two major roles: adhesion of cells to their substratum and transmission of signals from the external environment into the cell.
   2. The extracellular domains of integrins are large.
      a. Some α integrins have flattened, circular "propeller" domains.
      b. Other α integrins have I domains in addition to the propeller domain.
      c. The ß chains lack propellers, but all have I or I-like domains.
      d. The I domains contain ligand-binding sites specific for arginine-glycine-aspartic acid (RGD) sequences.
      e. The two subunits form a globular extracellular head connected to the membrane by elongated legs.
      f. The legs have a bent structure, the head faces the outer surface of the plasma membrane rather than the extracellular space.

g. The bent structure corresponds to the inactive state of the integrin. Activation of the integrin results in an upright conformation.
        3. RGD-containing drugs or toxins inhibit normal integrin function.
        4. Small cytoplasmic domains bind to cytoskeletal proteins.
    B. Focal contacts and hemidesmosomes:
        1. In vitro cells form stable attachments on contact with substrates and flatten over time. The cell membrane comes close to the plate in discrete places called focal contacts.
        2. In vivo, the tightest cell-substrate are mediated by hemidesmosomes, basal attachments of epithelial cells to the basement membrane.
        3. Contact between intracellular and extracellular structures occurs via integrins.

IV. Interactions of Cells with Other Cells
    A. Cells have surface-recognition sites that maintain organization within tissues.
    B. Selectins bind to specific carbohydrates on surfaces of cells.
        1. Extracellular domains contain multiple ligand-binding sites.
        2. Selectins mediate interactions between circulating leukocytes and vessel walls.
    C. Immunoglobulins and integrins:
        1. Most immunoglobulin superfamily (IgSF) proteins are involved in immunity.
        2. NCAM, VCAM, and L1 IgSF proteins mediate adhesion between nonimmune cells directly or through integrins.
    D. Cadherins are glycoproteins mediating Ca^{2+}-dependent cell adhesion.
        1. Cadherins join cells of similar type.
        2. Cadherins associate laterally forming dimers that are bridged by calcium ions in a "cell-adhesion zipper" model.
        3. The greater the number of interacting cadherins in a cluster, the greater the strength of adhesion between apposing cells.
        4. Expression of different cadherins during embryogenesis is responsible for cell rearrangement.
    E. Synaptic Junctions occur between axons and other nerves or specific cells.
        1. Protocadherins are cadherins that mediate specific synaptic connections.
        2. Cadherins have synaptic specificity. Cells in different parts of the brain and spinal cord contain different combinations of classic cadherins on their surfaces.
        3. Cadherins also connect pre- and post-synaptic cells.
    F. Adherens junctions and desmosomes: Anchoring cells to other cells.
        1. The adherens junctions in epithelia forms a belt around the cell's apical surface, called junctional complexes.
            a. Cell-to-cell contact in adherens junctions is cemented by calcium-dependant cadherin linkages.
            b. Catenins connect cadherins to intracellular cytoskeletal proteins.
        2. Desmosomes are tough, disc-shaped junctions between cells.
            a. Intracellular domains are linked to intermediate filaments.
            b. Intermediate filaments span the cell.
    G. The role of cell-adhesion receptors in transmembrane signaling:
        1. Binding of integrins and cadherins to extracellular ligands can transfer information into the cell.

V. The Human Perspective: The Role of Cell Adhesion in Inflammation and Metastasis
    A. Inflammation is a normal response to injury or infection but can be triggered inappropriately.
    B. Inflammation begins with the recruitment of leukocytes to the site of injury.
        1. Neutrophils attach to P-selectins on the walls of the venules.

2. Platelet activating factor is released by the venules.
    a. PAF alters the binding of neutrophilic integrins to IgSF (ICAMs).
    b. Neutrophils slip through the vessel walls (extravasation) into tissues.
C. Abnormal cell adhesion may be responsible for the spread (metastasis) of some cancers.

VI. Tight Junctions: Sealing the Extracellular Space
A. Tight junctions restrict the movement of material between epithelial cells.
B. Integral proteins form intermittent cell contacts near junctional complexes.
C. Continuous parallel strands of integral proteins encircle the cell.
    1. Occludins are found in tight junctions.
    2. Claudins occur with occludins and may be responsible for differences in tight junction permeability.
D. Tight junctions form the blood-brain barrier.

VII. Plasmodesmata: Mediating Intercellular Communication
A. Gap junctions connect cells via connexons, complexes of six connexin molecules.
B. Connexons have pores allowing diffusion of molecules less than 1000 daltons.
C. Gap junction intercellular communication (GJIC) integrates the activity of adjacent cells.
D. Compatibility differences between connexins may play improtant roles in either promoting or preventing communication between different types of cells in an organ.
E. Plasmodesmata are channels through cell walls, connecting adjacent cells in plants.

VIII. Cell Walls
A. Plant cell walls provide protection against abrasion, osmotic stress, and pathogens.
B. Microfibrils of cellulose form the fibrous component of the cell wall.
C. The matrix of the cell wall contains hemicellulose, pectins, and hydroxyproline-rich, proline-rich, and glycine-rich structural proteins.
D. Primary cell walls are flexible to allow growth after division.
E. Growth is facilitated by a class of pectins known as expansins.
F. Mature secondary cell walls are thicker and more rigid.

23

Key Figure

Figure 7.28. An overview of the types of interactions involving the cell surface.

Questions for Thought:

1. How do the proteins that participate in these different cell-to-cell contacts differ? What features do they have in common?
2. What do each of these kinds of adhesive structures do for the cell and the tissue in which they are found? What kinds of tissues might have an abundance of each respective kind of adhesive unit?
3. Which proteins could be involved in transmembrane signaling? Which proteins probably are not? We usually think of transmembrane signaling as a transfer of information from the outside of the cell to the inside. How might information go from inside the cell to the outside compartment?
4. Some metastatic cancers are characterized by nonfunctioning cell-adhesion proteins. Which proteins illustrated here might contribute to metastasis in mutant form?

Review Problems

Short Answer

1. Give the significance (not the definition) of the following terms or phrases. Say what they do or why they are important. For example:
    extracellular matrix: *plays a key role in determining the shape and activities of the cell and tissue.*

    a. glycocalyx

    b. basement membrane

    c. integrins

    d. RGD sequence

    e. matrix metalloproteinases

    f. hemidesmosomes

    g. cadherins

    h. tight junctions

    i. gap junctions

    j. plasmodesmata

2.   Compare and contrast the following:

   a.   glycocalyx vs. extracellular matrix

   b.   fibronectin vs. integrin

   c.   focal contacts vs. hemidesmosomes

   d.   tight junctions vs. gap junctions

   e.   gap junctions vs. plasmodesmata

Multiple Choice

1.   Which of the following is not an example of a type of extracellular matrix?
   a. basement membrane
   b. the cornea of the eye
   c. intermediate filaments
   d. bone

2.   Which of the following proteins of the ECM has a membrane bilayer-spanning domain?
   a. fibrillar collagen
   b. heparan sulfate proteoglycans
   c. fibronectin
   d. laminin

3.   Embryos injected with antibodies to fibronectin show inhibited movement of neural crest cells during development of the nervous system. These experiments show that:
   a. development of the neural crest involves the expression of antibody genes.
   b. developing neurons must synthesize fibronectin.
   c. fibronectin-antibody complexes form pathways for neural migration.
   d. neurons in embryos must transiently bind to fibronectin during migration.

4.   Although the protein families that comprise the ECM are very different, they have one of the following characteristics in common:
   a. They all have two or more distinct domains with specific binding sites.
   b. They are all derived from proteins of the immune system.
   c. They all have membrane-spanning domains.
   d. None of them are attached to the cell.

5. Which of the following is not a role of integrins?
   a. Integrins form a loose, hydrated gel in the ECM.
   b. Integrins anchor cells to the substrate.
   c. Integrins transmit signals to the intracellular compartment.
   d. Integrins participate in specialized cell-to-cell adhesive structures

6. Integrin is to a focal adhesion as ___?___ is to an adherens junction.
   a. integrin
   b. selectin
   c. fibronectin
   d. cadherin

7. The addition of a peptide with an RGD sequence would probably___?___ the binding of cultured cells to a fibronectin-coated dish.
   a. enhance
   b. inhibit
   c. have no effect on
   d. none of the above

8. The addition of a peptide with an RGD sequence would probably___?___ the binding of cultured cells to a collagen-coated or laminin-coated dish.
   a. enhance
   b. inhibit
   c. have no effect on
   d. none of the above

9. Focal adhesions are to cultured cells as ___?___ are to cells in vivo.
   a. plasmodesmata
   b. gap junctions
   c. cell walls
   d. hemidesmosomes

10. Actin filaments are to focal adhesions as___?___are to hemidesmosomes.
    a. myosin filaments
    b. keratin filaments
    c. microfilaments
    d. cadherins

11. If a laboratory mouse had a mutation rendering the gene that codes for the L-selectin protein nonfunctional, which of the following symptoms would the animal exhibit?
    a. blistering of the skin
    b. clotting deficiencies
    c. metastasis
    d. inability to fight infection in the tissues

12. Calcium-dependent adhesion is to the cadherins as calcium-independent adhesion is to the ?.
    a. IgSF proteins
    b. fibronectins
    c. integrins
    d. glycocalyx

13. To which category of proteins listed below are NCAM and VCAM most closely related?
    a. proteoglycans
    b. antibodies
    c. integrins
    d. selectins

14. Which of the following cell types might have the highest number of desmosomes?
    a. smooth muscle cells
    b. red blood cells
    c. epithelial cells of the skin
    d. nerve cells

15. How might anti-selectin antibodies act as anti-inflammatory drugs?
    a. Anti-selectin antibodies prevent activated endothelial cells from expressing selectin.
    b. Anti-selectin antibodies compete with carbohydrate ligands on neutrophil surfaces for selectin binding sites.
    c. Anti-selectin antibodies prevent neutrophils from transiently binding vessel walls.
    d. Anti-selectin antibodies prevent activated endothelial cells from producing platelet activating factor.

16. In going from the apical to the basal end of an epithelial cell, in what order do the various cell surface junctions appear?
    a. tight junction $\rightarrow$ adherens junction $\rightarrow$ desmosomes $\rightarrow$ hemidesmosomes
    b. desmosomes $\rightarrow$ hemidesmosomes $\rightarrow$ adherens junction $\rightarrow$ tight junction
    c. adherens junction $\rightarrow$ tight junction $\rightarrow$ desmosomes $\rightarrow$ hemidesmosomes
    d. tight junction $\rightarrow$ adherens junction $\rightarrow$ hemidesmosomes $\rightarrow$ desmosomes

17. Cells of the heart must contract in synchrony to produce an effective heart beat. The electrical signal telling each cell to contract must reach every cell of the heart at the same time. What kind of cell-to-cell junctions would you expect to find in heart tissue?
    a. gap junctions
    b. tight junctions
    c. desmosome
    d. adherens junctions

18. Plasmodesmata of plant cells have__?__in common with gap junctions in animal cells, but have__?__that gap junctions lack.
    a. connexin, connexon
    b. cytoplasmic continuity, desmotubules
    c. movement proteins, cytoplasmic continuity
    d. desmotubules, cell walls

19. Fluorscein that is injected into one plant cell___?___ show up in adjacent cells; DNA injected into one plant cell __?__ show up in adjacent cells.
    a. would not; would not
    b. would not; would
    c. would; would not
    d. would; would

20. That cancer cells often lack the gap junctions common to non-malignant cells of the same type is evidence that
    a. cancer is a disease of gap junctions.
    b. GJIC is an element of controlled cell growth.
    c. cancers all involve mutations in connexin genes.
    d. connexin is responsible for cell adhesion.

Problems and Essays

1. Your studies of the development of the insect nervous system have lead you to suspect that nerve cells that grow out of the central nervous system follow paths of extracellular laminin. These pathways, however, are rich in both laminin and collagen, and you cannot be sure that its the laminin and not the collagen that these cells are following. Design an experiment in vitro to distinguish between these two possibilities.

2. Indicate whether the following are extracellular (EX), intracellular (IN), or integral, membrane-spanning (MS) proteins:

   _____laminin
   _____collagen
   _____heparan sulfate proteoglycans (HSPG)
   _____connexin
   _____fibronectin
   _____keratins affiliated with hemidesmosomes
   _____integrin
   _____selectin
   _____NCAM
   _____catenins
   _____actin

3. The textbook suggests that unwanted blood clots may be treated by injecting synthetic RGD-containing peptides into the circulatory system. Name three other processes that might be influenced by this treatment.

4. In your lab you have four murine (mouse) cell lines growing in culture. Culture 1 is an epithelial line originated from embryonic glial cells. Culture 2 is a line of neural ectoderm, also started from embryonic cells. Culture 3 is derived from culture 1, but has been transfected with a gene for N-cadherin. Culture 4 is derived from culture 2, but has been transfected with a gene for E-cadherin. When you grow cells from each of the different cultures as a monolayer, then add radiolabeled, suspended cells from each of the cultures to each respective monolayer, you note that some labeled cells stick to some monolayers and not to others, and other labeled cells stick to different monolayers.

a. Indicate in the table which combinations of suspended and monolayered cells aggregate (+) and which do not (-).

b. A fifth culture of murine fibroblast cells, called L cells, express no cadherins. Add a line to your table indicating which combinations that involve culture 5 will form aggregates.

Cells in suspension

Cells in monolayer	1	2	3	4	5
1					
2					
3					
4					
5					

5. Metastatic cancer cells often exhibit changes in characteristics of cell surface proteins. Assume you are comparing the cell surface proteins of a particularly invasive cancer with those of normal cells. Describe how the level of each of these proteins differs in the cancer cells. Include a sentence describing how the change contributes to the malignant character of the cell.

cell surface protein	expression increased or decreased in cancer cell	contribution to cell's malignant character
laminin receptors		
fibronectin receptors		
integrins		
cadherins		
gap junctions		

6. The following are descriptions of some tissues with specialized functions as described. Indicate what types of cell-to-cell junctions probably hold these cells together, and why.
    a. Smooth muscle surrounds the digestive tract. The muscle must contract in synchrony to create the peristaltic movements that propel the contents through the intestine.

    b. The acini of the pancreas are clumps of cells that surround a lumen and secrete powerful digestive enzymes into the lumen destined for the intestine. Back-leakage of the secreted enzymes could cause severe cell damage to the tissues surrounding the acini.

    c. The smooth muscle of the uterus must withstand extreme stretching during pregnancy without tearing.

    d. The skin of a frog actively transports salts from the environment into the body, creating an osmotic and electrical gradient between the inside of the animal and the pond water.

    e. The longitudinal muscles and the circular muscles that surround the cavity of a sea anemone are complementary to one another. One of the sets of muscles contracts, the other relaxes. Each group acts as a single unit.

Web Explorations

Three hundred years ago, sailors on long voyages often suffered from scurvy, a horrible disease characterized by softening and bleeding of organs, tendons, skin, and gums. Scurvy often led to death. In the late 1700s, a young medical doctor named William Stark used himself as an experimental subject in studies of scurvy and its causes. By depriving himself of all but bread and water, Stark was able to give himself scurvy. He concluded that fresh fruits and vegetables could prevent the disease. Stark died at the young age of 29, probably from complications related to his scurvy experiments.

Using the World Wide Web, find the connection between scurvy, vitamin C (ascorbic acid) from fresh fruits and vegetables, and collagen of the extracellular matrix..
HINT: Search using key words "ascorbic acid," "collagen," and "scurvy." You might also search for key word "prolyl hydroxylase," the name of an enzyme that catalyzes hydroxylation of proline residues in collagen.

8

Cytoplasmic Membrane Systems: Structure, Function and Membrane Trafficking

Learning Objectives

When you have finished this chapter, you should be able to:

1. Explain the structural and functional relationship between the endoplasmic reticulum, Golgi complexes, lysosomes, and plasma membranes of eukaryotic cells.
   a. Describe the pathway of proteins destined for secretion, starting with synthesis and ending with exocytosis.
   b. Describe the pathway of materials taken up by the cell (including the various mechanisms of uptake), from the extracellular compartment to the final destination.
   c. Describe the trafficking of membranes in both secretory and cellular uptake processes.
2. Summarize the mechanisms whereby newly synthesized proteins are targeted and sorted for their appropriate destinations.
3. List the ways in which proteins can be modified and elaborated after they are synthesized; include the intracellular sites of these modifications.
4. Describe lysosomes and plant vacuoles. What do they have in common? Describe some human disorders associated with lysosomal malfunction.

Key Terms and Phrases

endomembrane system
secretory pathway
endocytic pathway
pulse-chase experiment
endoplasmic reticulum (ER)
cytoplasmic space
signal hypothesis
translocation
protein quality control
dolichol phosphate
trans Golgi network (TGN)
non-clathrin coated vesicles
anterograde movement
adaptin
secretory granules
turnover
vacuole
endocytosis
opsonin
low-density lipoprotein (LDL)

transport vesicle
constitutive secretion
sorting signal
cell fractionation
cisternal space (lumen)
signal sequence
SRP receptor
signal peptidase
glycosyltransferase
transitional elements
dictyosome
bulk flow
retrograde movement
clathrin
exocytosis
autophagy
tonoplast
phagocytosis
endosomes
triskelion

biosynthetic pathway
regulated secretion
autoradiography
microsomes
rough endoplasmic reticulum (RER)
smooth endoplasmic reticulum (SER)
signal recognition particle (SRP)
flippase
phospholipid exchange protein
Golgi complex
coat proteins (COPs)
adenosylation ribose factor (ARF)
t-SNAREs and v-SNAREs
N-linked and O-linked carbohydrates
lysosome
lysosomal storage disorder
receptor-mediated endocytosis (RME)
bulk-phase endocytosis
coated pits
high-density lipoprotein (HDL)

Reviewing the Chapter

I. Introduction
    A. Membranes divide the cytoplasm of eukaryotic cells into distinct compartments.

    B. Membrane-bound structures (organelles) are found in all eukaryotic cells.

II. An Overview of the Endomembrane System
    A. Most organelles are part of a dynamic system in which vesicles move between compartments.

    B. Biosynthetic pathways move proteins, carbohydrates and lipids within the cell.

    C. Secretory pathways discharge proteins from cells.
        1. Constitutive secretion is the continual, unregulated discharge of material from the cell.
        2. Regulated secretion is the discharge of products stored in cytoplasmic granules, in response to appropriate stimuli.

    D. Endocytic pathways move materials into cells.

    E. Sorting signals are recognized by receptors and target proteins to specific sites.

III. A Few Approaches to the Study of Cytomembranes
    A. Insights gained from autoradiography
        1. Autoradiography visualizes radiolabeled materials exposed to a photographic film.
        2. The site of protein synthesis in the cell can be visualized by incubating tissue with labeled amino acids.
        3. Brief incubation with labeled amino acids (pulse) and subsequent incubation with unlabeled amino acids (chase) allow the movements of proteins in the cell to be followed.

    B. Insights gained from the use of green fluorescent protein (GFP):
        1. Researchers have fused the DNA encoding a fluorescent protein from jellyfish with genes for secretory proteins.
        2. The GFP-secretory protein chimera allows them to follow secretory proteins in living cells.
        3. Viral proteins make good targets because infected cells produce copious viral proteins for secretion.

    C. Insights gained from the biochemical analysis of subcellular fractions:
        1. Subcellular fractionation is used to study the biochemistry of purified cell structures.
        2. Vesicles formed in cell homogenates have different properties allowing membranes from different organelles to be separated.
        3. Microsomes are heterogeneous mixtures of similar-sized vesicles, formed from membranes of the endoplasmic reticulum and Golgi complex.
        4. Microsomes retain activity during purification, allowing studies of function and composition.
        5. Proteins in cell fractions can be identified using proteomic technology which provides a comprehensive molecular portrait of any organelle

    D. Insights gained from the use of cell-free systems:
        1. Rough microsomes are artifactual vesicles of endoplasmic reticulum with attached ribosomes.
        2. Proteins synthesized on rough microsomes in cell-free systems ent4er the microsomal vesicle.
        3. Ribosomes experimentally stripped from rough microsomes still make proteins, but release them into the surrounding medium.
        4. Cell-free systems of synthetic liposomes are used to study the proteins responsible for budding and fusing.

E. Insights gained from the study of genetic mutants:
   1. Genetic mutants provide insights into the function of normal gene products.
   2. Isolation of proteins from yeast has led to the identification of homologous proteins in mammalian systems.
F. The dynamic activities of endomembrane systems are highly conserved despite the structural diversity of different cell types.

IV. The Endoplasmic Reticulum
A. Membranes of the endoplasmic reticulum (ER) enclose a luminal (cisternal) space.
B. The smooth Endoplasmic Reticulum (SER):
   1. Smooth ER (SER) is an interconnecting network of tubular membrane elements.
   2. Functions of the SER include:
      a. synthesis of steroids in endocrine cells.
      b. detoxification of organic compounds in liver cells.
      c. release of glucose 6-phosphate in liver cells.
      d. sequestration of Ca^{2+}.
C. Functions of the rough endoplasmic reticulum:
   1. Rough ER (RER) has ribosomes on the cytosolic side of continuous, flattened sacs (cisternae).
   2. The polarity of organelles in some secretory cells reflects the flow from the site of synthesis to the site of discharge.
   3. Synthesis of proteins on membrane-bound versus free ribosomes:
      a. Proteins synthesized on ribosomes of RER include secretory proteins, integral membrane proteins, and proteins of organelles.
      b. Proteins synthesized on "free" ribosomes include cytosolic proteins, peripheral membrane proteins, nuclear proteins, and proteins found in mitochondria and chloroplasts.
      c. The signal hypothesis explains how proteins are synthesized at different subcellular sites.
         (1) A signal sequence on the N-terminus attaches nascent secretory proteins to the ER.
         (2) The nascent polypeptide moves into the ER lumen cotranslationally.
         (3) A protein-lined pore forms a channel through which the peptide enters the ER lumen.
   4. Synthesis of secretory, lysosomal, or plant vacuolar proteins on membrane-bound ribosomes.
      a. Messenger RNA binds to free ribosomes in the cytosol.
      b. The signal sequence binds a signal recognition particle (SRP) which then attaches to an SRP receptor on the ER membrane.
      c. The SRP and the SRP-receptor must both bind to GTP in order to interact.
      d. Proteins undergoing translocation pass through a translocon channel which leads into the ER lumen.
      e. The release of the SRP requires a GTP-binding protein.
   5. Processing of newly synthesized proteins in the endoplasmic reticulum.
      a. Upon entering the ER lumen, the signal sequence is cleaved by signal peptidase and carbohydrates are added by oligosaccharyltransferase.
      b. The protein is folded with the help of chaperones, and disulfid bonds are added by protein disulfide isomerase (PDI).
      c. Improperly folded proteins are moved into the cytosol and destroyed in a process called quality control.
   6. Synthesis of integral membrane proteins on membrane-bound ribosomes.

       a. Integral membrane proteins have hydrophobic sequence(s) that stop transfer into the ER lumen.

       b. The translocon orients transmembrane sequences properly.

D. Membrane biosynthesis in the ER:
1. Lipids are inserted into existing membranes.
2. Asymmetry of proteins and lipids is maintained during membrane trafficking.
3. Some integral membrane proteins are synthesized on ribosome of the ER and translocated into the ER membrane cotranslationally.
4. Integral membrane proteins have hydrophobic stop-transfer sequences that cause them to bind to the inner surface of the translocon.

E. Synthesis of membrane lipids: Lipids inserted into the outer leaflet of the ER are modified by:
1. conversion of one type of phospholipid to another type.
2. selective inclusion in nascent vesicles.
3. phospholipid transfer proteins that move lipids between membranes.

F. Glycosylation in the rough endoplasmic reticulum:
1. Carbohydrates are added by glycosyltransferases to almost all proteins produced on membrane-bound ribosomes.
2. Sugars are attached in an ordered sequence.
3. The core carbohydrate chain is assembled on a lipid carrier, dolichol phosphate.
4. The core carbohydrate is transferred to the polypeptide by oligosaccharyltransferase and is modified en route to the Golgi complex.
5. Misfolded proteins are tagged with a terminal glucose and transferred to a chaparone to be refolded.
6. Congenital Diseases of Glycosylation (CDGs) are cause by mutations that lead to partial disruption of the glycosylation pathway.

G. Mechanisms that ensure the destruction of misfolded proteins:
1. Misfolded proteins are not destroyed in the ER; they are transported to the cytosol by "reverse translocation".
2. The proteins are destroyed in the cytosol by proteasomes.
3. Accumulation of misfolded proteins in the ER triggers an unfolded protein response (UPR).
4. Sensors on the ER are kept inactive by molecular chaperones, particularly BiP.
5. BiP molecules get 'tied up' when they interact with misfolded proteins rendering they incapable of inhibiting the sensors.
6. Activated sensors send signals to the nucleus and cytosol promoting the expression of hundreds of different genes that encode the following proteins:
       a. ER-based molecular chaperones that help proteins reach the native state
       b. Proteins involved in the transport of misfolded proteins out of the ER
       c. Proteins involved in the selective destruction of abnormal proteins

G. From the ER to the golgi complex: The first step in vesicular transport:
1. Membrane and luminal proteins move to the tips of the RER.
2. Transitional elements (i.e., RER tips) form the first transport vesicles.
3. ER-derived transport vesicles fuse to form a large ERGIC (endoplasmic reticulum Golgi intermediate complex) on the way to the Golgi.

V. The Golgi Complex
A. Golgi are stacks of flattened cisternae lacking ribosomes.

B. The cis face of the Golgi faces the ER; the trans face is on the opposite end of the stack.
C. The Golgi networks are processing and sorting stations where proteins are modified, segregated and then shipped in different directions.
   1. The cis Golgi network (CGN) sorts proteins for the ER or the next Golgi station.
   2. The trans Golgi network (TGN) sorts proteins to various cellular destinations.
   3. These membranous elements of the Golgi are supported mechanically by a peripheral membrane skeleton.
D. Glycosylation in the Golgi Complex:
   1. Synthesis of complex polysaccharides takes place in the Golgi.
   2. Oligosaccharides attach to glycoproteins and glycolipids in the Golgi.
E. The movement of materials through the Golgi complex: Two views
   1. In the cisternal maturation model, cisternae formed on the cis face are transient and physically move to the trans face.
   2. The more recent vesicular transport model holds that material moves through permanent Golgi stacks via vesicles.
   3. The current model is similar to the cisternal maturation model, but also includes retrograde movement via vesicles.

VI. Types of Vesicle Transport and Their Functions

A. Materials are carried between compartments by various protein-coated vesicles.
   1. COPII-coated vesicles move materials from the ER to the Golgi.
     a. A small GTP-binding protein called Sar plays a regulatory role in assembly and disassembly of the vesicle coat.
     b. Sar-GTP binds to the ER; Sar-GDP dissociates from the ER.
   2. COPI-coated vesicles move materials from the Golgi to the ER.
     a. Adenosylation ribose factor (ARF), a GTP-binding protein, is required for vesicle transfer between cisternae.
     b. COPI coated vesicles may select specific cargo.
   3. Clathrin-coated vesicles move materials from the trans Golgi to endosomes, lysosomes, and plant vacuoles.
B. Organelles retain resident molecules and retrieve "escaped" molecules.
   1. Proteins contain unique retrieval signals that ensure their return to the proper organelle.
   2. Specific receptors capture and return "runaway" molecules to their original organelle.
   3. Anterograde and retrograde movements through the Golgi are possible.
C. Beyond the Golgi Complex: Sorting proteins at the TGN:
   1.. Lysosomal proteins are tagged with phosphate groups in the Golgi cisternae.
     a. These proteins are captured by mannose 6-phosphate receptors in the TGN.
     b. The cytosolic domains of the receptors bind adaptin molecules and direct the vesicles to lysosomes.
   2. Nonlysosomal proteins also have sorting sequences
     a. Secretory proteins aggregate in dense granules that bud from the TGN.
     b. Integral plasma membrane proteins have signal sequences in the cytoplasmic domains.
D. Targeting vesicles to a particular compartment:
   1. Movement through the cytoplasm may be directed by microtubules.
   2. Vesicles are tethered to target compartments, a process mediated by GTP-binding proteins called Rabs.
   3. SNARES mediate docking between target and vesicle.
     a. v-SNAREs occur in vesicles.

b. t-SNAREs occur on target sites.
c. SNAREs are a protein family including VAMP-like proteins and syntaxin.
4. Target and vesicle membranes fuse, a process mediated by intracellular proteins.
E. Exocytosis
1. Cells engaged in extensive secretion have numerous secretory vesicles.
2. The process of exocytosis is triggered by local increases in Ca^{2+} concentration.
3. Fusion proteins found on the vesicle and membrane mediate the formation of fusion pores.
4. Vesicles may or may not become absorbed into the plasma membrane.

VII. Lysosomes

A. Lysosomes contain acid hydrolases that can digest every kind of biological molecule.
B. Internal proton concentration is kept high by H^+-ATPases.
C. Glycosylated proteins, lgp-A and lgp-B, may protect the lysosome from self-digestion.
D. Lysosomes are identified by the presence of the enzyme acid phosphatase.
E. Lysosomes are involved in two major cell functions– phagocytosis and autophagy.
F. Primary lysosomes fuse with either phagocytic or autophagic vesicles, forming residual bodies that either undergo exocytosis or are retained in the cell as lipofuscin granules.
G. The Human Perspective: Disorders resulting from defects in lysosomal function:
1. Lysosomal malfunction can have profound effects on human health.
2. Silicosis and asbestosis result from the lysosomal uptake of indigestible fibers.
3. Lysosomal storage disorders result from the absence of specific lysosomal enzymes allowing undigested material to accumulate.
4. Most hereditary lysosomal disorders can be diagnosed prenatally.

VIII. The Human Perspective: Disorders Resulting from Defects in Lysosomal Function

A. A break or leak in the membrane of lysosomes releases digestive enzymes into the cell which damages the surrounding tissues.
B. A mannose 6-phosphate residue in the lysosomal enzymes act as an address for delivering these proteins to the lysosomes.
C. Lysosomal storage disorders (accumulation of nondegraded substrate in the lysosome) occur because of a deficiency in lysosomal enzymes.
1. Symptoms of these disorders depend on the degree of enzyme dysfunction.
2. Tay-Sachs disease results from a deficiency of the enzyme (-N-hexosaminidase A) whose function is to degrade gangliosides, a major component of brain cell membranes.
3. Enzyme replacement therapy, gene therapy, and bone marrow transplantation are currently being tested as treatment for these disorders.

IX. Plant Cell Vacuoles

A. Plant cells contain large vacuoles for the storage of compounds.
B. Transport systems in the tonoplast (vacuole membrane) accumulate material in the vacuole, attracting water by osmosis; this in turn generates turgor pressure, providing support for the cell.
C. Plant vacuoles, like lysosomes of animal cells, are an end point for biosynthetic pathways and may contain acid hydrolases.

X. The Endocytic Pathway: Moving Membrane and Materials into the Cell Interior

A. Endocytosis is the cellular uptake of material dissolved or suspended in fluid.
1. Bulk-phase endocytosis does not require surface membrane recognition.

2. Receptor-mediated endocytosis (RME) follows the binding of substances to membrane receptors.
   a. Ligand-bound receptors collect in specialized regions of the membrane forming coated pits.
   b. Clathrin-coated regions invaginate into the cytoplasm, forming coated vesicles.
   c. Clathrin contains three chains that form a triskelion capable of changing from a flat sheet to a cage like scaffold surrounding the vesicle.
   d. The formation of a clathrin-coated endocytic vesicle requires an additional GTP-binding protein called dynamin.
   e. Adaptor complexes between the clathrin lattice and the membrane interact with specific signals in the receptors being internalized.
   f. The vesicle loses its clathrin coat and may fuse with an early endosome. The receptors may be recycled to the plasma membrane.
3. The endocytic pathway is divided into the early endosomes and late endosomes.
   a. Materials in the early endosomes are sorted and integral membrane proteins are shipped back to the membrane and other dissolved materials and bound ligands are sent to the late endosomes.
   b. Molecules that reach the late endosomes are moved to lysosomes.

B. LDLs and cholesterol metabolism:
   1. Cholesterol is transported in the blood in lipoprotein complexes.
   2. LDLs are taken up by RME and delivered to lysosomes, releasing the cholesterol for use by the cells.
   3. HDLs transport cholesterol from tissues to the liver, for elimination.
   4. HDL is associated with lowering cholesterol levels, whereas LDL is associated with high blood cholesterol and the formation of atherosclerosis.

C. Phagocytosis is for uptake of large particles
   1. The plasma membrane engulfs a particle and pinches off to form a phagosome.
   2. The phagosome fuses with a lysosome, where the contents are digested in a phagolysosome.
   3. Phagocytosis is for eating in some unicellular organisms, and for protection in most mammaliam phagocytic cells.

XI. <u>Posttranslational Uptake of Proteins by Peroxisomes, Mitochondria, and Chloroplasts</u>
   A. Proteins targeted to nuclei, mitochondria, chloroplasts and peroxisomes contain signal sequences that serve as addresses.
   B. Uptake of proteins into peroxisomes is mediated by a peroxisomal targeting signal (PTS) on the protein and cytosolic PTS receptors.
   C. Uptake of proteins into the mitochondria:
      1. Many mitochondrial proteins are encoded by nuclear genes and synthesized in the cytosol.
      2. A sequence in the N-terminus directs transport of polypeptides from the cytoplasm into the matrix.
      3. Proteins enter the mitochondria at specialized sites where the outer and inner membranes make contact.
         a. The TOM complex acts as a receptor and channel across the OMM.
         b. The TIM complex acts as a receptor and channel across the IMM.
      4. Translocation is powered by the voltage gradient.
      5. A cytosolic chaperone (Hsp70) unfolds the polypeptide for transport through the membrane; a matrix chaperone (Hsp60) refolds the protein inside the mitochondrion.

6. Some proteins enter the matrix and are then threaded back either into the inner membrane or completely through it into the intermembrane space.
7. Proteins from mitochondrial genes are made on ribosomes attached to IMM.
D. Uptake of proteins into chloroplasts
1. Most chloroplast proteins are imported from the cytosol.
2. Outer and inner envelope membranes have Toc and Tic complexes to aid in transport.
3. Chaperones unfold proteins in the cytosol destined for chloroplasts.
4. A removable transit peptide targets proteins to the specfic chloroplast. Subcompartments.
5. Proteins with a thylakoid transfer domain target proteins for the thylakoids.

XII. **Experimental Pathways: Receptor-Mediated Endocytosis**
A. The mechanisms of endocytosis at specialized coated pits was first proposed by Roth and Porter to explain the uptake of yolk protein by oocytes.
B. The detailed structure of the coated vesicles was revealed by electron microscopy.
C. Barbara Pearse identified clathrin as the predominant protein by SDS-PAGE electrophoresis of proteins from purified fractions of coated vesicles.
D. Cells from individuals with familial hypercholesterolemia (FH) are unable to regulate cholesterol synthesis in response to the presence of LDL.
E. Brown and Goldstein showed that in affected individuals, the primary defect is the inability of the LDL receptor to initiate RME.
F. A key tyrosine residue may be essential for the localization of ligand bound receptors in coated pits.

Key Figure

Figure 8.2. An overview of biosynthetic/
secretory and endocytic pathways that unite
endomembranes into a dynamic,
interconnected network.

Questions for Thought:

1. Starting with the rough endoplasmic
   reticulum, trace a patch of membrane
   out to the cell's outer boundary, then
   back again to the Golgi complex. What
   lipid modifications may take place
   during this migration?

2. If you were doing histochemistry on
   the cell illustrated here, which
   membrane compartments would test
   positive for carbohydrates?

3. Which vesicles illustrated here might
   be surrounded by clathrin?

4. The exact mechanism of regulated
   secretion is only partially understood.
   We know that calcium and membrane
   proteins on both the secretory vesicle
   and the plasma membrane are
   involved. Can you propose a
   mechanism that includes these
   components? How might you test your
   model?

Review Problems

1. Give the significance (not the definition) of the following terms or phrases. Say what they do or why they are important. For example:
   endoplasmic reticulum: *delineates an intracellular compartment that functions in the sorting of newly synthesized proteins and provides a surface for the synthesis of many proteins.*

   a. regulated secretion

   b. cell fractionation

   c. signal sequence

   d. *trans* Golgi network

   e. retrograde movement of Golgi vesicles

   f. phosphorylated mannose residues on TGN glycoproteins

   g. lysosomes

   h. acid hydrolases

   i. receptor-mediated endocytosis (RME)

   j. coated pits

2. Compare and contrast the following:

   a. pulse vs. chase

   b. protein synthesis on membrane-bound vs. free ribosomes

   c. synthesis of N-linked vs. O-linked Oligosaccharides

   d. autophagy vs. phagocytosis

   e. LDL vs. HDL

Multiple Choice

1. Cytoplasmic membrane systems are characteristic of what types of cells?
   a. prokaryotic cells
   b. eukaryotic cells
   c. bacterial cells
   d. all cells

2. Pancreatic cells that produce and release digestive enzymes engage in what type of secretion?
   a. regulated secretion
   b. constitutive secretion
   c. transport secretion
   d. none of the above

3. Which of the following cell types would be the best model system for studying secretory granules?
   a. muscle cells
   b. yeast cells
   c. epithelial cells
   d. pancreatic cells

4. You are interested in studying the asymmetric (coming from just one cell side) release of digestive enzymes from epithelial cells that line the digestive tract. Which of the following techniques would be best for these studies?
   a. cell fractionation
   b. homogenization
   c. autoradiography
   d. all of the above

5. Which of the following explains why microsomes can't be seen in cells viewed with the electron microscope?
    a. They are far too small.
    b. They are artifacts of homogenization and centrifugation.
    c. They are transparent to electrons.
    d. They actually can be seen in electron micrographs of cells.

6. Which of the following proteins would not be found in the smooth endoplasmic reticulum?
    a. Ca^{2+}-pumping enzymes
    b. cytochrome P_{450}
    c. glucose 6-phosphatase
    d. signal peptidase

7. Which of the following groups of proteins probably lack a signal sequence?
    a. acid hydrolase enzymes synthesized in macrophage cells
    b. glycolytic enzymes synthesized in liver cells
    c. polypeptide hormones synthesized in endocrine cells
    d. antibody hormones synthesized in plasma cells

8. Which of the following is not an essential component of the complex that directs a nascent protein into the lumen of the RER?
    a. protein disulfide isomerase
    b. SRP
    c. SRP receptor
    d. GTP-binding protein

9. If you compared the proteins in a *cis* Golgi compartment with those in a *trans* Golgi compartment, you would find:
    a. the proteins in the two compartments are identical.
    b. the proteins in the *cis* compartment are glycosylated and contain modified amino acids, whereas those in the *trans* compartment are not modified.
    c. the proteins in the *cis* compartment are glycosylated, whereas those in the *trans* compartment are glycosylated and contain modified amino acids.
    d. the proteins of the *cis* compartment are shorter than those of the *trans* compartment.

10. You are tracing the path of a secretory protein from its synthesis to its export from a cell. You have added radioactive amino acids to a culture of cells, and then measured the amount of radioactivity that shows up in the proteins of each of the following cell fractions at different times after the addition. List the order in which the proteins of these fractions first exhibit radioactivity:
    >   I   secretory vesicles
    >   II   Golgi complex
    >   III rough ER
    >   IV   smooth ER
    >   V   nucleus
    a. III → II → I → out of the cell
    b. III → II → IV → V → out of the cell
    c. V → III → II → I → out of the cell
    d. IV → III → II → I → out of the cell

11. Of the following processes that occur during protein trafficking, which involve GTP?
    a. protein translocation across the ER membrane
    b. disassembly of the COP coat on a transport vesicle
    c. fusion of non-clathrin coated vesicles with target membrane
    d. all of the above

12. You have labeled the lipids on a patch of rough ER membrane with a fluorescent probe. After a few minutes, the probe shows up in the membranes of the *cis* Golgi. Now you treat the cells with the drug brefeldin A. Where might the fluorescent probe show up next?
    a. *trans* Golgi network
    b. endoplasmic reticulum
    c. plasma membrane
    d. secretory vesicles

13. v-SNARE proteins are to transport vesicles as KDEL sequences are to__?__.
    a. t-SNARE proteins
    b. escaped ER proteins
    c. ERGIC vesicles
    d. coat proteins

14. Which type of vesicle of the *trans* Golgi network would be most likely to carry hormones destined for regulated secretion?
    a. lysosomal vesicles
    b. clathrin-coated vesicles
    c. non-clathrin-coated vesicles
    d. all of the above

15. COPII-coated vesicles are to anterograde movement as__?__are to retrograde movement.
    a. ERGIC-coated vesicles
    b. clathrin-coated vesicles
    c. t-SNARE-coated vesicles
    d. COPI coated vesicles.

16. Of the following, which would be least likely to be found in lysosomes?
    a. acid hydrolase enzymes
    b. a half-digested mitochondrion
    c. nucleic acids
    d. a high concentration of protons

17. In plant cells, the structures most analogous to lysosomes are called __?__.
    a. vacuoles
    b. peroxisomes
    c. digestive vesicles
    d. lysosomes

18. Uptake of low-density lipoproteins (LDL) occurs by__?__, whereas retrieval of plasma membrane after extensive secretory activity occurs by__?__.
    a. receptor-mediated endocytosis, bulk-phase endocytosis
    b. bulk-phase endocytosis, receptor-mediated endocytosis
    c. receptor-mediated endocytosis, receptor-mediated endocytosis
    d. bulk-phase endocytosis, bulk-phase endocytosis

19. Which of the following are not associated with coated pits?
    a. TGN
    b. triskelion
    c. adaptor complexes
    d. integral membrane receptors

20. Which cells of the body would probably be the richest in HDL receptors?
    a. stomach cells
    b. liver cells
    c. muscle cells
    d. epithelial cells

Problems and Essays

1. For each of the following proteins, indicate whether it is synthesized on "free" ribosomes (FR) or on the rough endoplasmic reticulum (RER).

    _____fibronectin

    _____cytoskeletal proteins such as actin

    _____acid hydrolases

    _____collagen

    _____triskelion

    _____the enzymes of glycolysis

    _____adaptin

    _____LDL receptor

    _____signal peptidase

    _____protein disulfide isomerase

    _____mitochondrial proteins coded for by nuclear DNA

2. Jamieson and Palade used the pulse-chase method to follow the fate of radiolabeled amino acids, from protein synthesis to secretion. Using what you know about endocytosis, use the same method (just on paper) to follow the fate of radiolabeled amino acids that are incorporated into the proteins of a bacterial cell that is engulfed by a macrophage.

3. Insulin is synthesized in pancreatic islet cells first as a precursor called proinsulin; then it is transformed into mature insulin by a proteolytic cleavage of a short peptide, via a protease enzyme. Your laboratory has antibodies specific for both proinsulin (anti-P) and insulin (anti-I). Both antibodies can be complexed with electron-dense gold particles and viewed in electron micrographs as dark spots. You have treated one specimen of thinly sliced pancreatic tissue with anti-P and another with anti-I, and viewed both in the electron microscope. The + signs on the following table indicate where the black dots appear in your micrographs:

	anti-P	anti-I	anti-clathrin
mature secretory vesicles	-	+	
trans Golgi network	+	-	
cis and medial Golgi cisternae	+	-	

a. Where does proteolytic conversion of proinsulin take place?

b. Assume you have treated a third specimen with anti-clathrin antibodies. On the table, indicate which, if any, vesicles would stain positive for clathrin?

4. A certain integral protein of the plasma membrane has two transmembrane domains in which both the C-terminus and the N-terminus are directed toward the cell exterior in the mature protein. Where are the C- and N-termini located when the protein is newly translocated into the RER membrane? Into the Golgi? Into secretory vesicles?

5. Protein synthesis can be studied in cell-free systems using free amino acids, mRNA and an extract of wheat germ, which contains ribosomes and all of the cofactors necessary for translation. You are using such a system in your lab to study translocation of nascent proteins. For each of the experimental conditions described below, indicate whether the protein is completely synthesized, whether the mature protein has a signal sequence, and whether it is translocated across a membrane.

Experimental Conditions	completely synthesized?	signal sequence?	translocated across a membrane?
a. mRNA codes for a cytosolic protein; no SRP present; no SRP receptor present; no microsomes present			
b. mRNA codes for a secretory protein; no SRP present; no SRP receptor present; no microsomes present			
c. mRNA codes for a secretory protein; exogenous SRP added; no SRP receptor present; no microsomes present			
d. mRNA codes for a secretory protein; exogenous SRP added; free (not membrane bound) SPR receptor added; no microsomes			
e. mRNA codes for a secretory protein; exogenous SRP added; exogenous SRP receptor added; microsomes added			

6. When an endosome containing LDL bound to LDL receptors fuses with a lysosome, the drop in pH decreases the affinity of the particle for the receptor, and both the lipid and the carrier proteins are degraded in the fused vesicle. Iron, like cholesterol, is transported in the blood as a complex with the transport protein transferrin. It is the iron-transferrin complex, called ferrotransferrin, that serves as the ligand for the transferrin receptor on cell membranes. Unlike the response of LDL, a decrease in pH does not lower the affinity of the ligand for the receptor. It does, however, lower the affinity of iron for the carrier protein transferrin. Propose a sequence of events in cells that take up ferrotransferrin, and follow the fates of both transferrin and the transferrin receptor following iron uptake.

Web Explorations

Until the early 1970s, asbestos was used widely in construction, especially in insulation material. Asbestos provides strength, heat insulation, and fireproofing-- all desirable characteristics for building materials. Unfortunately, inhaled asbestos fibers cause chronic lung disease, gradual destruction of lung tissues, and often death. Asbestosis, the disease resulting from inhalation of asbestos, is characterized by the presence of minute asbestos fibers in the lysosomes of lung cells.

Using the World Wide Web, learn more about asbestosis. Who is most at risk for this disease? What are the effects of particle accumulation on lung tissue? What happens to the particles after a cell that has ingested them dies? Why do the symptoms of asbestosis often appear 30 or 40 years after exposure to asbestos?

9

The Cytoskeleton and Cell Motility

Learning Objectives

When you have finished this chapter, you should be able to:

1. Describe the three types of cytoskeletal structures.
    a. For each structure, list the features of the component proteins and explain how the proteins interact.
    b. Explain how each structure is assembled and disassembled, and how these processes are regulated.
2. Define "molecular motor." Name the three families of motor proteins.
    a. State where each type of motor protein is likely to occur and with which cytoskeletal structures each is affiliated.
    b. Explain how motor proteins convert chemical energy into cellular movement.
3. Name the cellular activities that involve the cytoskeleton, including those that affect movement and those that do not.
4. Describe the organization of cytoskeletal and accessory proteins in skeletal muscle. How does this arrangement relate to muscle contraction? How is contraction regulated?
5. Understand the techniques used to study the cytoskeleton and cellular motility. Appreciate why these techniques are well suited to these types of studies.

Key Terms and Phrases

cytoskeleton	intermediate filaments (IFs)	microfilaments
microtubules	fluorescence microscopy	tubulin
actin	video microscopy	myosin
kinesin	ciliary and cytoplasmic dynein	protofilament
plus and minus ends	microtubule-associated protein (MAP)	centrosome
centriole	microtubule-organizing center (MTOC)	pericentriolar material
basal body	cilia	flagella
axoneme	muscle fiber	myofibrils
sarcomere	thin and thick filaments	striations
transverse (T) tubules	neuromuscular junction	sarcoplasmic reticulum (SR)
actin-binding proteins	excitation-contraction coupling	pseudopodia
lamellipodia	contact inhibition of movement	growth cone

Reviewing the Chapter

I. <u>Introduction and Overview of the Major Functions of the Cytoskeleton</u>
    A. The cytoskeleton is a dynamic structure with many roles.
        1. It serves as a scaffold, providing structural support and maintaining cell shape.
        2. It serves as an internal framework, organizing organelles within the cell.
        3. It assists in movement of materials within the cell and cellular locomotion.

4. It provides anchoring sites for mRNA.
5. It serves as a signal transducer.
6. It anchors mRNA during translation.
7 It is an essential component of the cell's division machinery.
B. The cytoskeleton is a network of three filamentous structures: microtubules, microfilaments, and intermediate filaments.

II. The Study of the Cytoskeleton
A. Fluorescent microscopy of live cells locates fluorescently-labeled proteins.
B. Video microscopy and in vitro motility assays enable observation of intracellular movements.
C. The electron microscope can be used to visualize the cytoskeleton of a treated cell.
D. The use of cells with altered gene expression shows functions of specific proteins.
    1. Knock-out animals lack a specific gene whose function can be inferred from the animal's abnormalities.
    2. Dominant negative mutant cells, engineered to overexpress a mutant protein, can illustrate the normal role of the protein.

III. Microtubules
A. Structure and composition:
    1. Microtubules are hollow cylindrical structures found in the cytoskeleton, mitotic spindle, cilia, and flagella.
    2. The microtubule is a polymer made up of globular tubulin subunits arranged in longitudinal rows called protofilaments.
    3. Most microtubules contain 13 protofilaments.
    4. Heterodimers of a- and ß-tubulin subunits are assembled into tubules with plus (fast-growing) and minus (slow-growing) ends.
B. Microtubule-associated proteins (MAPs):
    1. A MAP contains a globular "head" that attaches to a microtubule and a filamentous "tail" that extends away from the microtubule's surface.
    2. MAPs form crossbridges, organizing microtubules into bundles, and also stabilize, alter the rigidity of, or influence the assembly of microtubules.
    3. MAPs are regulated by phosphorylation of specific amino acid residues.
C. Microtubules as structural supports and organizers:
    1. The distribution of microtubules corresponds to the shape of the cell.
    2. Nerve axons have microtubules oriented parallel to the long axis of the axon.
        a. In mature axons, microtubules function in axonal transport of vesicles.
        b. During embryonic development, microtubules play a key role in axon growth.
    3. Microtubules in plant cells organize cellulose-synthesizing enzymes that produce the cell wall.
    4. Microtubules play a role in the location of organelles.
D. Microtubules as agents of intracellular motility:
    1. Microtubules facilitate the movement of vesicles traveling between compartments.
    2. Axonal transport:
        a. Neurotransmitters are synthesized in the cell body and released from synapses at the ends of axons.
        b. Both anterograde transport (away from the cell body) and retrograde transport (toward the cell body) involve microtubules.
E. Motor proteins that traverse the microtubular cytoskeleton
    1. Molecular motors convert chemical energy of ATP into mechanical energy.
    2. Molecular motors move by stepwise, conformational changes corresponding to a mechanical cycle.
    3. Three categories of molecular motors have been described.

      a. Kinesins and dyneins move along tracks of microtubules.

      b. Myosins move along microfilament tracks.

F. Kinesins:

  1. Kinesin is a member of a superfamily of more than 50 proteins, called kinesin-like proteins (KLPs).

  2. Kinesins have a pair of force-generating globular heads and a fan-shaped, cargo-binding tail.

      a. The motor domains of kinesin and kinesinlike proteins are similar.

      b. Tail sequences differ reflecting the variety of cargo these proteins can haul.

  3. Most kinesins are plus end-directed microtubular motors responsible for anterograde transport.

      a. The Ncd kinesins in Drosophila are minus end-directed.

      b. Ncd differs from other kinesins in the neck region of the protein.

  4. Kinesin molecules move along a single protofilament of a microtubule at a velocity proportional to the ATP concentration.

  5. Kinesin movement is processive and operates by a alternating, "hand-over-hand" mechanism.

  6. Kinesinlike proteins, found in all types of eukaryotic cells, are implicated in movement of membrane-bound cargo towards the cell's plasma membrane.

G. Cytoplasmic Dynein:

  1. Dyneins are associated with cilia and flagella but are also found in the cytoplasm.

  2. Cytoplasmic dyneins are large proteins with globular, force-generating heads.

  3. Dyneins move toward the minus ends of microtubules and are involved in chromosomal movement during mitosis, minus end-directed movement of vesicles, and retrograde axonal transport.

H. Microtubule organizing centers (MTOCs):

  1. Centrosomes have two centrioles surrounded by pericentriolar material.

      a. Centrioles have nine fibrils, each with three microtubules.

      b. Centrioles are found in pairs oriented at right angles to each other.

      c. Microtubules grow from the pericentriolar material of a centrosome.

      d. Centrosomes form the poles from which the mitotic spindle emerges.

  2. Basal bodies and other MTOCs:

      a. Basal bodies are the MTOCs for cilia and flagella.

      b. MTOCs control the number of microtubules, their polarity, the number of protofilaments in their walls and the time and location of their assembly.

      c. The protein g-tubulin occurs in all MTOCs and is essential for microtubule assembly.

      d. Only the γ-tubulin of the heterodimer can bind to a ring of subunits creating the polarity of the microtubule.

I. The dynamic properties of microtubules:

  1. Microtubules assemble and disassemble readily (dynamic instability).

  2. Microtubules are polymers formed by the noncovalent association of dimeric building blocks that constantly rearrange and reassemble.

  3. Tubulin dimers are added or removed at the plus end, away from the MTOC.

J. The study of microtubules in vitro:

  1. GTP on the ß-subunit of the tubulin dimer is hydrolyzed during polymerization.

      a. Dimers bound to GTP have a higher affinity for one another than those bound to GDP.

             b. GTP hydrolysis provides a mechanism for regulation of assembly and disassembly.

      2. Temperature, Ca^{2+} concentration, and MAPs influence microtubule stability.

K. The study of microtubules in vivo:

      1. Cells injected with fluorescent labeled tubulin illustrate dynamic instability of microtubules.

      2. Cells respond to changing conditions by rapidly remodeling microtubules.

L. Cilia and flagella:

      1 Cilia and flagella have similar structures but different motions.

      2. Cilia are relatively short and occur in large numbers on cell surfaces.

      3. Flagella, longer and fewer in number, are used primarily for motility.

M. The structure of cilia and flagella:

      1. The central core (axoneme) of both cilia and flagella contain microtubules in a 9 + 2 arrangement.

            a. The peripheral doublets contain both an A and a B tubule.

            b. Radial spokes connect the A tubules to the central tubules.

      2. Cilia and flagella emerge from basal bodies.

      3. Basal bodies contain nine triplet microtubules and no central microtubules.

      4. Embryonic ciliary action may contribute to body symmetry.

N. The dynein arms:

      1. The machinery for ciliary and flagellar function resides in the axoneme.

      2. Ciliary dynein is needed for ATP hydrolysis which supplies the energy for locomotion.

O. The mechanism of ciliary and flagellar locomotion:

      1. Swinging cross-bridges generate the forces for ciliary or flagellar movement.

      2. The dynein arm of an A tubule binds to a neighboring B tubule and forces a conformational change that slides the tubules past each other.

      3. Sliding alternates from one side of the axoneme to the other causing that part of the cilium or flagellum to bend first in one direction and then the opposite direction.

P. Ciliary and flagellar locomotion is regulated with the dynein arm activity which is controlled by the central pair of microtubules and the radial spokes.

IV. The Role of Cilia in Development and Disease

A. *Situs inversus* is a syndrome in which people exhibit reversed left-right body symmetry.

B. One cause of *situs inversus* is mutations in genes encoding ciliary proteins.

D. Nonmotile cilia on other cells detect properties of surrounding fluids.

V. Intermediate Filaments

A. Intermediate filaments (IFs) are a heterogeneous group of proteins, each with a central helix flanked by globular domains.

B. IFs assemble by several mechanisms.

      1. The basic unit of assembly is a tetramer formed by two antiparallel dimers.

      2. Both the tetramer and the IF lack polarity.

C. IFs resist tensile forces and IFs containing keratin form the protective barrier of skin.

D. Assembly and disassembly of IFs are controlled by phosphorylation and dephosphorylation.

E. IFs include neurofilaments, the major component of the structural framework supporting neurons.

F. IFs may provide specialized mechanical support in certain cell types.

      1. Defective IFs may lead to extreme fragility.

      2. Over expression of IFs in neurons can lead to degenerative diseases.

VI. Microfilaments

A. Microfilaments are made of actin and are involved in cell motility.
B. Using ATP, globular actin polymerizes to form actin filaments (F actin).
C. Actin is a major contractile protein in muscle but is present in all cell types.
D. Microfilament assembly and disassembly:
   1. G actin is added at the plus end and removed from the minus end, causing treadmilling.
   2. The microfilament cytoskeleton is organized by controlling the equilibrium between actin polymers and monomers.
E. Actin polymerization can act as a force-generating mechanism in some cells.
F. Myosin: The molecular motor for actin filaments:
   1. All myosin share a characteristic motor (head) domain but the tail domains are highly divergent and are divided into two groups of conventional (type II) and unconventional myosins.
   2. Conventional (type II) myosins generate forces in muscle and some non-muscle processes.
      a. Myosin II has a tail composed of two heavy chains, two necks, and two globular heads that contain the catalytic site.
      b. All of the machinery required for motor activity is contained in a single head domain.
      c. Myosin forms bipolar filaments with tails all oriented toward the center.
   3. Unconventional myosins have only a single head and are unable to assemble into filaments in vitro.
      a. Myosin I is capable of actin-based processes but not cell division.
      b. Myosin V moves along actin without disconnecting; i.e., processively.
      c. Several unconventional myosins are associated with membranous organelles.
      d. Myosin VIIa is found in hair cells of the ear.
      e. Myosin VI moves toward the minus end of actin.

VII. Muscle Contractility

A. Skeletal muscle fibers are large, multinucleate cells packed with myofibrils that contain repeating arrays of sarcomeres.
B. The sliding filament model of muscle contraction:
   1. Thick and thin filaments in the sarcomere overlap.
   2. Sarcomeres extend between Z lines.
   3. I bands contain only thin filaments, the H zone contains only thick filaments, and the A band contains both thick and thin filaments.
   4. Skeletal muscle functions by shortening fiber length.
   5 Three classes of filaments are found in muscle cells: actin, myosin, and titin.
      a. Titin filaments provide tension in resting muscles and support myosin filaments.
      b. Tropomyosin molecules occupy the groove between the two actin molecules.
      c. Heterotrimeric troponin molecules are spaced evenly along thin filaments.
      d. Myosin head project laterally from the thick filaments.
   6. During contraction, the myosin heads form cross-bridges with the actin filaments.
      a. The myosin heads bend, sliding the thin actin filaments over the thick filament.
      b.. The energy release by ATP hydrolysis induces a small conformational change within the head.
      c. This is amplified by the swinging movement of the elongated neck.

>   d. The movement causes the actin filaments to slide a much greater distance than would otherwise be possible.
> 7. Energy is provided by ATPase activity in the myosin head.
>   a. ATP binds to the head inducing the dissociation of the cross-bridge from the actin filament..
>   b. The activated myosin attaches to the actin initiating the power stroke.
>   c. The release of bound ADP is followed by binding of another ATP.
>   d. The absence of ATP prevents dissociation of the cross-bridges and is the basis for rigor mortis, the condition following death.
>
> C. Excitation-contraction coupling: Muscle fibers are innervated by branches of a single motor neuron form motor units.
>   1. The contact between nerve and muscle occurs at the neuromuscular junction.
>   2. Linking the nerve impulse to the shortening of the sarcomere is called excitation-contraction coupling.
>   3. Action potentials are carried into the cell interior by transverse (T) tubules.
>   4. T tubules terminate near the sarcoplasmic reticulum (SR), which sequesters Ca^{2+}.
>   5. In the relaxed state, cytoplasmic Ca^{2+} levels are low. Action potentials open calcium channels in the SR, releasing Ca^{2+}.
>   6. The binding of Ca^{2+} to troponin C causes a conformational change, shifting the tropomyosin and exposing the myosin binding site.
>   7. Ca^{2+} is removed from the cytoplasm by the Ca^{2+}-ATPase activity in the SR, causing tropomyosin to hide myosin-binding sites.

VIII. Nonmuscle Motility

> A. Actin-binding proteins organize actin filaments into functional assemblies.
>   1. Monomer-sequestering proteins (profilin and thymosin) bind to G actin and prevent polymerization.
>   2. End-capping proteins regulate the length of actin filaments.
>   3. Cross-linking proteins link together two or more separate actin filaments.
>   a. Rod-shaped cross-linking proteins promote the formation of networks that have the properties of elastic gels.
>   b. Globular actin-binding proteins bundle actin filaments into tight parallel arrays which are found in microvilli and stereocilia.
>   4. Filament-severing proteins shorten filaments, decreasing cytoplasmic viscosity.
>   5. Actin-filament depolymerizing proteins (cofilin family of proteins) stimulate the breakdown of actin into monomer units.
>   6. Membrane-binding proteins link contractile proteins to the plasma membrane.
>
> B. Examples of nonmuscle motility and contractility:
>   1. Actin filaments and myosin motors are responsible for cytokinesis, phagocytosis, cytoplasmic streaming, vesicle trafficking, movement of integral proteins in membranes, cell locomotion and more.
>   2. Cell-substratum interactions and the role of stress fibers:
>   a. Stress fibers attach to the substrate at focal contacts and contain bundles of actin filaments.
>   b. Stress fibers contain other proteins found in muscle cells.
>   c. Stress fibers contract isometrically, creating tension between cell and substrate.
>   3. Actin polymerization as a force-generating mechanism:
>   a. The bacterium Listeria moves through host cells by causing actin to polymerize behind it.
>   b. Actin polymerization can also move intracellular vesicles, or whole cells.

4. Cell locomotion:
   a. Cells lacking cilia or flagella move by "crawling" over the substrate.
   b. Cultured fibroblasts crawl by forming lamellipodia.
   c. The protrusion of lamellipodia involves the organization of actin filaments.
   d. Force generation in lamellipodia occurs by the addition of actin monomers to cortical filaments and/or the movement of myosin motors along cortical filaments.
   e. Cell movement requires both the formation and breakdown of microfilaments and adhesions, occurring simultaneously in different regions of the cells.
5. Axonal outgrowth occurs in embryonic nerve cells.
   a. Highly motile growth cones contain several structures filled with actin.
   b. Growth cones follow defined markers during embryonic development.
6. Changes in cell shape during embryonic development are due to the cytoskeleton.

IX. **Experimental Pathways on the Web: The Molecular Motor That Drives Fast Axonal Transport**
   A. A technique called video-enhanced contrast, differential interference microscopy, or AVEC-DIC, allowed high-resolution images of living cells, facilitating the study of axonal transport.
   B. Allen showed that vesicular movements along linear elements are both anterograde and retrograde.
   C. In 1985, Vale, Schnapp, Reese and Sheetz showed that vesicles of different sizes move along linear elements at equal rates, indicating that a single motor drives fast axonal transport.
      1. Organelles attached to single filaments move in both directions, even passing one another, suggesting that each filament has more than one track.
      2. The filaments were the diameter of microtubules and bound immunofluorescent antibodies to tubulin.
   D. Vale's group reconstituted the axonal transport system from separate components.
      1. The addition of a supernatant fraction (S2) to the mixture of tubulin and organelles promoted transport in only the anterograde direction.
         a. The motor protein kinesin was isolated from S2.
         b. Kinesin was identified as a plus end-directed motor.
      2. A retrograde motor was identified using affinity chromatography to remove kinesin from the extracts.
      3. Drug sensitivity suggests that the retrograde motor may be a dyneinlike protein.
   E. A kinesin receptor molecule on transported vesicles, called kinectin, is part of a larger family of vesicle-transport proteins.

Key Figure

Figure 9.1*a*. An overview of the structure and functions of the cytoskeleton.

(*a*) **Epithelial cell**

Questions for Thought:

1. What role does is the cytoskeleton playing in the asymmetric structure of this cell? Compare this diagram to figure 8.11 in Chapter 8. What role do you think the cytoskeleton plays in the polarity of the cell in that figure?
2. What motor protein is illustrated in the circular subdiagram labeled "motor protein?" How do you know?
3. In addition to where they are illustrated in this figure, where else in the cell can each of these cytoskeletal elements be found?
4. What changes would occur in the cytoskeleton of this cell if it were to undergo cell division?

Review Problems

1. Give the significance (not the definition) of the following terms or phrases. Say what they do or why they are important. For example:
    molecular motors: *convert chemical energy in the form of ATP to mechanical energy in the form of movement.*

    a. overlapping functions of cytoskeletal proteins

    b. microtubule-organizing centers (MTOCs)

    c. the high metabolic cost of microtubule polymerization

    d. γ–tubulin

    e. intermediate filaments occur only in animal cells

    f. high degree of conservation in the structure of actin proteins

    g. experiments of A. Huxley, Niedergerke, H. Huxley, and Hanson

    h. actin-binding proteins

    i. contractile stress fibers

    j. contact inhibition of movement

2. Compare and contrast the following:

   a. microtubules vs. microfilaments

   b. cytoplasmic dynein vs. kinesin

   c. basal bodies vs. centrioles

   d. muscle fibers vs. myofibrils

   e. pseudopodia vs. lamellipodia

Multiple Choice

1. The vertebrate skeleton is to bones as the cytoskeleton is to ___?___.
   a. calcium
   b. tubulin, actin and dynein
   c. microtubules, microfilaments, and intermediate filaments
   d. scaffolding

2. The movement of a motor protein along a cytoskeletal filament is most analogous to:
   a. a locomotive rolling along a railroad track.
   b. a truck driving along the highway.
   c. a person on crutches walking down a sidewalk.
   d. a jet plane in flight.

3. Which of the following is *not* typical of the "cargo" moved by molecular motors?
   a. ATP
   b. secretory vesicles
   c. cytoskeletal filaments
   d. organelles

4. Differences in microtubular stability and function between different cells (or different locations within a cell) may be due to which of the following?
   a. the expression of different tubulin isoforms
   b. the number of protofilaments composing a single microtubule
   c. the presence of different MAPs
   d. all of the above

5. If you treated cells with a drug that interferes with microtubules, such as colchicine, which of the following would result?
   a. Cell shape would be disrupted.
   b. Mitosis and meiosis would not occur.
   c. The intracellular location of organelles would be disrupted.
   d. All of the above would result.

6. Which of the molecular motor proteins is associated with microtubules?
   a. kinesins
   b. dyneins
   c. myosins
   d. both (a) and (b)

7. Which of the following cytoskeletal components is not found in all eukaryotic cells?
   a. dynein
   b. intermediate filaments
   c. microtubules
   d. kinesinlike proteins

8. Dynein is to retrograde movement as ___?___ is to anterograde movement.
   a. kinesin
   b. myosin
   c. ATP
   d. tubulin

9. Which of the following is not an example of a microtubule-organizing center (MTOC)?
   a. a centrosome
   b. a basal body
   c. a tubulin-GTP dimer
   d. a spindle pole body

10. First you dissolve the membrane from an intact flagellum using the detergent Triton X-100. Next you soak the axoneme in a solution containing EDTA, which removes the Mg^{2+}. What remains of the axoneme after these treatments?
    a. peripheral tubules only
    b. peripheral tubules and central tubules, but no side arms or ATPase activity
    c. peripheral tubules, central tubules, side arms, and ATPase activity
    d. peripheral tubules, central tubules, side arms, ATPase activity, and a membrane

11. In what way are intermediate filaments similar to both microtubules and microfilaments?
    a. IFs are a chemically heterogeneous group of structures.
    b. IFs assemble and disassemble under changing cellular conditions.
    c. IFs have been identified in animal cells but not in other eukaryotes.
    d. IFs assemble into structures of varying thickness via several different pathways.

12. What kind of information can be obtained by injecting cells with fragments of heavy meromyosin (HMM) or S1 subfragments of HMM?
    a. identification of actin-containing filaments and their respective polarities
    b. identification of actin-containing filaments but not their respective polarities
    c. identification of myosin-containing filaments and their respective polarities
    d. identification of myosin-containing filaments but not their respective polarities

13. Which of the following cell types would be the best source of actin?
    a. vertebrate skeletal muscle cells
    b. asymmetric cells, such as those lining the digestive tract
    c. plant cells in the process of active cell division
    d. skin cells

14. Which of the following is essential for treadmilling to occur down the length of an actin filament?
    a. incubation with HMM
    b. asymmetry in the rates of polymerization and depolymerization occurring at the two respective ends of the filament
    c. stability of the actin filament
    d. all of the above

15. Which types of cell motility can occur in the presence of actin alone (without myosin)?
    a. skeletal muscle contraction
    b. cytokinesis
    c. the acrosomal reaction just prior to the fertilization of an egg by a sperm
    d. none (All types of motility using actin also require myosin.)

16. In an analogy between skeletal muscle motility and a railroad, if actin is the railroad tracks and ATP is the diesel fuel, what is the locomotive?
    a. myosin
    b. titin
    c. troponin
    d. tropomyosin

17. If you compared under the microscope a contracted sarcomere to a relaxed sarcomere, which of the following regions would not differ in width?
    a. the A band
    b. the I band
    c. the H zone
    d. the entire sarcomere

18. During the course of a muscle twitch (a contraction due to a single action potential), when is the free calcium concentration of the muscle cytoplasm highest?
    a. milliseconds before the T tubule undergoes an action potential
    b. after the T tubule undergoes an action potential but before the muscle generates tension
    c. at the height of the tension generated by the muscle
    d. at the end of the twitch, after the muscle has relaxed

19. The sarcoplasmic reticulum must have integral membrane proteins that can:
    a. release and pump Ca^{2+}.
    b. bind to tropomyosin and troponin.
    c. undergo action potentials.
    d. contract.

20. The diversity of roles played by microtubules largely results from the diversity of tubulin isoforms that make up the filaments. What accounts for the diversity of roles played by microfilaments?
    a. actin isoforms
    b. variations in the manner in which G actin aggregates into F actin
    c. differences in the myosins that interact with actin
    d. the variety of actin-binding proteins

Problems and Essays

1. You are observing a microtubule using video-enhanced microscopy. At various time intervals, you measure the length of both ends (A and B) of the microtubule using a fixed reference point. You have plotted your data below:

a. Which is faster, assembly or disassembly? How can you tell?

b. Which line represents the changes occurring at the plus end? Why?

c. What is the name of the phenomenon illustrated here?

d. What are some naturally occurring intracellular events that may correspond to the peaks and valleys of these lines?

e. At the point labeled M, draw what would happen to this microtubule if the cell in which it resides entered mitosis.

2. Indicate which of the following processes are directly affected by colchicine (C), taxol (T), cytochalasin (CHL), and/or nonhydrolyzable ATP analogs, such as AMP-PNP :

_____axonal transport

_____mitotic spindle formation

_____mitotic spindle disassembly

_____acrosomal reaction

_____phagocytosis

_____cytokinesis

3. Panel A is a schematic diagram of a cross section of a sarcomere from skeletal muscle.

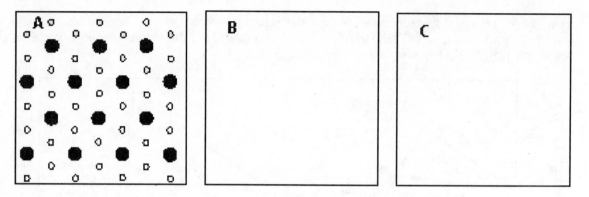

   a. From which region of the sarcomere is this cross section derived?

   b. In panel B, depict an H zone cross section, and in panel C, an I band cross section.

   c. How might the cross section in panel A differ between contracted and relaxed muscle?

4. Colchicine, an alkaloid from the saffron plant, binds to tubulin dimers and prevents their polymerization. If applied to actively dividing cells, colchicine causes the disappearance of the mitotic spindle, thus blocking cell division. A different drug, taxol, has just the opposite effect. In the presence of taxol, microtubules polymerize or are stabilized. This, it turns out, is just as deleterious to the dividing cell as is colchicine. Can you explain this based on what you know about the roles of microtubules in cells?

5. The isolated axoneme of a cilium can be gently treated with proteases that disrupt the nexin links between microtubules and the radial spokes between peripheral and central microtubules. When these axonemes are exposed to ATP, the fiber appears to rapidly elongate, up to nine times its original length. Explain.

6. A skeletal muscle cell has three distinct membrane systems, each with its own complement of integral membrane proteins. Indicate which membrane– plasma membrane (PM), sarcoplasmic reticulum (SR), or no membrane (NONE)– would be richest in each of the following proteins.

_____tropomyosin

_____acetylcholine receptor

_____Ca²⁺-ATPase

_____titin

_____Ca²⁺-release channels

7. Fill in the blanks in the following table:

	Ciliary Movement	Skeletal Muscle Movement
Filament Type		
Motor Protein		
Fuel Molecule		
Name of the Functional Unit of Movement		

Web Explorations

In 1971, a compound isolated from the bark of the Pacific yew tree was implicated as a possible cancer therapy. The compound, called paclitaxel, was shown to inhibit cell division in tumor cells by promoting the assembly of and inhibiting the disassembly of microtubules. Several problems, however, slowed wide scale development and testing of the compound in cancer patients. Paclitaxel was in short supply and expensive to produce. Nonetheless, in 1991 Bristol-Meyers Squibb started full scale development of the drug under the trade name of Taxol. Today Taxol is used to treat a wide range of cancers, including advanced ovarian and breast cancer, certain lung cancers (non-small cell) in people who cannot have surgery or radiation therapy, and AIDS-related Kaposi's sarcoma.

Using the World Wide Web, learn more about the source of Taxol and the costs of isolating the compound. How is Taxol produced today? What are the side effects of the drug and how can you explain them on the basis of what you know about the cytoskeleton?

10 The Nature of the Gene and the Genome

Learning Objectives

When you have finished this chapter, you should be able to:

1. Understand Mendel's rules of inheritance and their significance for genetics.
2. Define the linkage group, and explain the reasons for incomplete linkage.
3. Give an overview of the discoveries that led Watson and Crick to formulate their structure for DNA. Describe the structure of DNA.
4. Describe the different conformations that the DNA double helix can assume.
5. Understand genome complexity and the methods used to determine it. Define three classes of DNA sequences.
6. Know the definition of transposons, their effect on gene expression, and their role in evolution.
7. Define a restriction enzyme. Give examples of restriction enzymes, and describe their role in the development of modern molecular biology.
8. Give examples of how the techniques of modern molecular biology are being used in medicine and technology.

Key Terms and Phrases

gene
law of independent assortment
linkage group
crossing over
polytene chromosomes
purine
Rosalind Franklin
double helix
complementarity
A form of DNA
negative supercoiling
genome
melting temperature ™
nucleic acid hybridization
satellite DNA
in situ hybridization
LINEs
deletion
Barbara McClintock
transposon
polymorphic genes
restriction endonuclease
restriction map
Frederick Griffith

allele
homologous chromosomes
T. H. Morgan
genetic recombination
nucleotide
base composition analysis
Maurice Wilkins
major groove
template
Z-DNA
positive supercoiling
complexity of the genome
DNA renaturation
highly repeated sequences
microsatellite DNA
moderately repeated sequences
nonrepeated sequences
pseudogenes
transposition
transposase
genetic map
restriction fragment length polymorphisms (RFLPs)
gene therapy
Avery, MacLeod, and McCarty

law of segregation
bivalent
mutation
locus
pyrimidine
Chargaff's rules
Watson and Crick
minor groove
B form of DNA
DNA supercoiling
topoisomerase
DNA denaturation
DNA reannealing
tandem sequence
minisatellite DNA
SINEs
duplication
mobile genetic element
transposable element
Alu sequence
physical map
Friedrich Miescher
Hershey and Chase

Reviewing the Chapter

I. **Introduction**
   A. Hereditary factors reside on chromosomes.
   B. They consist of DNA.

II. **The Concept of a Gene as a Unit of Inheritance**
   A. Gregor Mendel's work became the foundation for the science of genetics.
   B. Mendel established the laws of inheritance based on his studies of pea plants.
       1. Characteristics of organisms are governed by units of inheritance called genes.
           a. Each trait is controlled by two genes called alleles.
           b. Alleles can be identical or nonidentical.
           c. When they are nonidentical, the dominant allele masks the recessive allele.
       2. A reproductive cell called a gamete contains one gene for each trait.
           a. Somatic cells arise by the union of male and female gametes.
           b. In somatic cells, two alleles controlling each trait are inherited; one from each parent.
       3. According to the law of segregation, the pairs of genes become separated during gamete formation.
       4. According to the law of independent assortment, genes controlling different traits segregate independently of each other.

III. **Chromosomes: The Physical Carriers of Genes**
   A. The discovery of chromosomes:
       1. Chromosomes were first observed in dividing cells, using the light microscope.
       2. During cell division, chromosomes are divided equally between the two daughter cells.
       3. Chromosomes are doubled prior to cell division.
   B. The chromosomes as the carriers of genetic information:
       1. Chromosomes are present as pairs of homologues.
       2. During meiosis, homologous chromosomes first associate and form a bivalent; they then separate into different cells.
       3. Behavior of pairs of chromosomes in meiosis correlates with Mendel's laws of inheritance.
   C. The chromosome as a linkage group:
       1. Genes that are on the same chromosome do not assort independently.
       2. All of the genes on a chromosome form a linkage group.
   D. Genetic analysis in *Drosophila*:
       1. T. H. Morgan was the first to use fruit flies in genetic research.
       2. Mutant fruit flies were isolated and became the primary tool for geneticists.
       3. Mutation was recognized as the raw material of evolution.
       4. Correlation between the four linkage groups and the four pairs of homologous chromosomes in *Drosophila* cells supported the idea of genes residing directly on chromosomes.
   E. Crossing over and recombination:
       1. Linkage between alleles on the same chromosome is incomplete.
       2. Crossing over (or genetic recombination) produces offspring with new combinations of genetic traits.
           a. The percentage of recombination between two genes is constant.
           b. The percentage of recombination between different pairs of genes can be different.

3. The position of genes along the chromosome (their locus) is fixed, and can be mapped.
4. The frequency of recombination provides a measure of distance between genes.
5. Recombination frequency increases with distance between genes.
F. Mutagenesis and giant chromosomes:
1. Exposure to a sublethal dose of x-rays increases spontaneous mutation frequency.
2. Cells in the larval salivary gland of *Drosophila* have giant polytene chromosomes.
3. Polytene chromosomes display a specific pattern when they have been stained; individual genes can be correlated with specific bands.
4. Gene expression can be directly visualized by observing the "puffs" in polytene chromosomes.

IV. The Chemical Nature of the Gene
A. DNA is the genetic material in all organisms.
B. The structure of DNA:
1. The earliest information on DNA structure was derived from biochemical studies and x-ray diffraction analyses.
2. The nucleotide is the basic building block of DNA.
    a. A nucleotide consists of a phosphate, a sugar, and either a purine or pyrimidine nitrogenous base.
    b. Nucleotides are linked into nucleic acid polymers:
        (1) Sugar and phosphates are linked by 3', 5'-phosphodiester bonds.
        (2) The nitrogenous bases project out like stacked shelves.
    c. Nucleic acids have a polarized structure; the ends are marked 5' and 3'.
3. Chargaff established rules by base composition analysis.
    a. Number of adenines (A) equals number of thymines (T).
    b. Number of cytosines © equals number of guanines (G).
    c. (A) + (T) (G) + (C)
C. The Watson-Crick proposal:
1. The DNA molecule is a double helix.
    a. The DNA molecule is composed of two chains of nucleotides.
    b. The two chains spiral around each other forming a pair of right-handed helices.
    c. The two chains are antiparallel (they run in opposite directions).
    d. The sugar and phosphate backbone is located on the outside of the molecule.
    e. The bases are situated inside the double helix; they are stacked perpendicular to the long axis.
    f. Two DNA chains are held together by hydrogen bonds between each base of one chain and a base on the other chain.
    g. The double helix is 2 nm wide.
    h. Pyrimidines are always paired with purines.
    i. Only A-T and G-C pairs fit into the double helix structure.
    j. From the outside, the DNA molecule has a major groove and a minor groove.
    k. The double helix takes a turn every 10 residues.
    l. The two chains are complementary to each other.
2. The importance of the Watson-Crick proposal:
    a. The Watson-Crick model of DNA as the genetic material accounts for its functions:

(1) Storage of genetic information.
(2) Self-duplication and inheritance.
(3) Expression of the genetic message.
3. DNA supercoiling:
    a. DNA can be either positively (overwound) or negatively (underwound) supercoiled.
    b. Topoisomerases change the level of supercoiling.

V. The Structure of the Genome

A. The genome of a cell or organism is its unique complement of genetic information.
B. The complexity of the genome:
  1. The complexity of the genome refers to the variety of genomic DNA sequences and the numbers of copies of those sequences.
  2. DNA denaturation is the separation of the two DNA strands; the melting temperature of the double helix depends on the G-C content.
  3. DNA renaturation is the reassociation of single strands into a stable double helix.
    a. Reannealing led to the development of a methodology called nucleic acid hybridization.
  4. For bacteria and viruses, the rate of renaturation is directly proportional to the size of the genome.
  5. The complexity of eukaryotic genomes: Reannealing curves show three classes of DNA.
    a. Highly repeated DNA sequences represent 10 % of the genome and do not code for gene products.
      (1) Satellite DNA has been located at the centromere using in situ hybridization.
      (2) Minisatellite DNAs occupy less of the genome than satellite DNAs and are used in DNA fingerprinting.
      (3) Microsatellite DNAs have been implicated in genetic disorders.
    b. Moderately repeated DNA sequences make up 20-80% of the total DNA.
      (1) Repeated DNA sequences with coding functions include the genes for histones, transfer RNAs, and ribosomal RNAs.
      (2) Repeated DNAs that lack coding functions include SINEs and LINEs.
    c. Nonrepeated DNA sequences code for the majority of proteins.
C. The Human Perspective: Diseases that Result from Expansion of Trinucleotide Repeats:
  1. Dynamic mutations occur in genes containing a repeating unit of three nucleotides.
  2. The mutant alleles are highly unstable and the number of repeating units tends to increase as the gene passes from parent to offspring.
  3. When the number significantly increases, individuals develop serious diseases.
    a. neurodegenerative disorders such as Huntington's disease
    b diseases affecting the brain and numerous parts of the body such as fragile X syndrome

VI. The Stability of the Genome

A. Information content of the genome can change rapidly.
B. Whole Genome Duplication (Polyploidization)
  1. Polyploidization occurs when offspring receive more than two sets of chromosomes from their parents.

      a. This can occur when hybrids are formed from closely related parents.

      b. It can also occur when duplicated chromosomes are not separated in embryonic cells.

  2. In polyploids that survive, extra copies of genes have evolutionary potential.

C. Duplication and modification of DNA sequences:

  1. Genetic exchange between misaligned homologous chromosomes can create either a duplication or a deletion of DNA sequences.

  2. Duplication played a major role in the evolution of multigene families, such as globin genes.

D. Evolution of the globin genes:

  1. The globin gene family includes the peptides of the hemoglobin tetramer, myoglobin, and the plant protein, leghemoglobin.

  2. Ancestral forms give rise to recent forms by duplication, gene fusion, and divergence.

  3. Some sequences, called pseudogenes, resemble globin genes but are nonfunctional.

  4. Pseudogenes are nonfunctional members of a multigene family.

  5. Evolution of DNA sequences can be observed in cultured cells in the laboratory.

E. "Jumping Genes" and the dynamic nature of the genome:

  1. Genetic elements are capable of moving.

      a. This genetic rearrangement is called transposition and the moving genes are transposons.

      b. Only certain sequences can act as transposons, but transposons insert into target sites randomly.

      c. Transposition requires the enzyme transposase coded within the transposable element itself, which facilitates insertion of transposable elements into the target site.

      d. Bacterial transposition occurs by replication of the transposable element, followed by insertion.

      e. In eukaryotic cells, transposition can occur either by replication via an RNA intermediate or by excision followed by reinsertion.

      f. Transposons that jump into a protein-coding gene are responsible for genetic diseases such as hemophilia.

  2. The role of mobile genetic elements in evolution:

      a. Some moderately repeated sequences in human DNA (Alu and L1) are transposable elements.

      b. According to one school of thought, transposable elements have no function and thus are genetic parasites.

      c. Transposable elements may play important evolutionary roles:

         (1) Some regulatory sequences are derived from transposons.

         (2) Transposons rearrange the genome, possibly linking the coding regions for different domains, thus creating multidomain proteins into composite genes.

         (3) Some transposable elements have given rise to new genes.

  3. Transposition has had an important effect on evolution of genomes.

VII. Sequencing Genomes: The Genetic Basis of Being Human

A. Genomes of hundreds of organisms have been sequenced.

B. In 2001, the entire human genome was published.

  1. The human genome contains about 30,000 protein-encoding genes.

  2. About 60% of protein-encoding regions may code for more than one protein.

      a. Alternate splicing of RNA may account for several proteins from one gene.

　　　　b. Posttranslational modifications also account for different protein functions.
　　3. It is difficult to find the exact number of human genes in the genome.
　C. Comparative genomics: If it's conserved, it must be important.
　　1. DNA that is similar among related organisms is thought to be important, even if the role is still unknown.
　　2. Some important DNA in humans may be of recent origin.
　　　　a. For example, the FOXP2 gene in humans differs slightly from that in chimpanzees.
　　　　b. Humans with mutated FOXP2 genes have language disorders.
　D. The Human Perspective: Application of Genomic Analysis to Medicine
　　1. The Human Genome Project helps identify genes responsible for complex diseases.
　　2. Researchers correlate genetic polymorphisms with incidence of various diseases.
　　　　a. SNPs are single nucleotide polymorphisms that represent the most common differences between individuals.
　　　　b. SNPs may play an important role in susceptibility to disease, or at least act as genetic markers for susceptibility.
　　　　c. SNP analysis may also indicate a person's responsiveness to a drug.
　　　　d. SNPs can be inherited in blocks, or groups, called haplotypes.
　　　　　　(1) Haplotype maps (HapMaps) are based on common haplotypes.
　　　　　　(2) HapMaps may lead to associations between disease and different haplotypes.
　　3. Polymorphisms can tell us something about our ancestral past.

VIII. Experimental Pathways: The Chemical Nature of the Gene
　A. The nature of the gene was discovered through a series of initially unrelated studies.
　B. Friedrich Miescher first identified "nuclein" in white blood cell extracts and in salmon sperm.
　C. The tetranucleotide theory, proposed by Phoebus Levene, stated that DNA was composed of a monotonous repetition of four nucleotide building blocks and could not be the genetic material.
　D. Transformation of bacterial cells was first discovered by Frederick Griffith in experiments with pneumococcus bacteria.
　E. The role of DNA as a transforming agent was shown in bacterial cell studies by Avery, MacLeod, and McCarty, and in bacteriophage studies by Hershey and Chase.

Key Figure

Figure 10.25 A pathway for the evolution of globin genes.

Questions for Thought:

1. What different kinds of genetic events happen that give rise to gene families? Can you think of examples of these processes in other contexts?

2. What evidence is there indicating that the ancestral globin gene existed even before the lineages of plants and animals diverged?

3. How did some members of the globin gene family end up on different chromosomes than others?

4. In humans, the δ-globin gene makes very little mRNA, and in gibbons, the δ-globin gene makes no mRNA. What might be happening to silence this gene? Why is gene silencing not harmful to humans or gibbons?

Review Problems

1. Give the significance (not the definition) of the following terms or phrases. Say what they do or why they are important. For example:
    nucleotide sequence in DNA: *contains the genetic information.*

    a. crossing over

    b. polytene chromosomes

    c. topoisomerases

    d. T. H. Morgan's experiments with *Drosophila*

    e. DNA supercoiling

    f. Barbara McClintock's experiments with maize

    g. single nucleotide polymorphisms (SNPs)

    h. Avery, MacLeod, and McCarty's experiments with pneumococcus

2.   Compare and contrast the following:

a.   gene vs. allele

b.   Mendel's law of segregation vs. his law of independent assortment

c.   genomic complexity of prokaryotes vs. genomic complexity of eukaryotes

d.   number of human protein coding sequences vs. number of human proteins

e.   conserved vs. variable sequences

Multiple Choice

1.   Why were the traits that Mendel chose to study in the pea plant considered lucky choices?
    a. They were the only traits governed by genes.
    b. None of the traits he chose acted as if they were part of the same linkage group.
    c. All the traits he chose could be found on the same plant.
    d. They all obeyed the law of segregation.

2.   Alleles are:
    a. different forms of the gene for the same trait.
    b. genes for different traits.
    c. always identical.
    d. inherited in pairs, both members of which come from the same parent.

3.   A brown-eyed man has one allele for brown eyes and one for blue eyes. Half his sperm carry the allele for brown eyes, and half carry the allele for blue eyes. This is an illustration of:
    a. Mendel's law of independent segregation.
    b. Mendel's law of independent assortment.
    c. Mendel's law of segregation.
    d. None of the above.

4. Which of the following observations provided evidence that Mendel's "units" were located on the chromosomes?
    a. Chromosomes occur as homologous pairs.
    b. The number of linkage groups and chromosomes is always equal.
    c. One member of each homologous pair goes to each gamete during meiosis.
    d. All of the above.

5. Given that alleles are located on chromosomes, and whole chromosomes are passed on to gametes, why don't offspring get all the alleles on a parental chromosome?
    a. Genetic recombination is a result of gene duplication during evolution.
    b. Genetic recombination is a result of mutation.
    c. Genetic recombination is a result of crossover between members of homologous pairs.
    d. All traits on a given chromosome are always found together in offspring.

6. Watson and Crick used all of the following information in deducing the structure of the genetic material except:
    a. Avery, MacLeod, and McCarty's identification of DNA as the "transforming principle."
    b. x-ray diffraction studies by Astbury, Franklin, Pauling and Wilkins.
    c. Chargaff's rules of DNA base composition.
    d. microscopic examination of DNA structure.

7. Given the structure of DNA as described by Watson and Crick, all of the following must be true of DNA except:
    a. The 3' end of one of the strands must be aligned with the 5' end of the other strand.
    b. The total number of purine bases must equal the total number of pyrimidine bases.
    c. The sequence of bases on one strand must be identical to those on the other strand.
    d. The sequence of bases in the genome of one organism must be different from the sequence of bases in the genome of a different organism.

8. The structure of different DNAs can vary in all of the following ways except:
    a. The helix may be right-handed or left-handed.
    b. The sequence of nucleotide bases can be different.
    c. The molecule can be fully hydrated as in the B form, or drier as in the A form.
    d. The bases can be attached to the sugars or the phosphates of the backbone.

Use the following diagram to answer questions (9), (10) and (11).

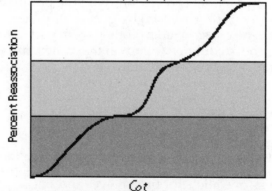

9. Which portion of the curve represents renaturation of the nonhistone coding sequences of the DNA?
    a. A
    b. B
    c. C
    d. cannot determine from this curve

10. Which portion of the curve represents renaturation of the most repetitive DNA?
    a. A
    b. B
    c. C
    d. cannot determine from this curve

11. What is probably the source of the DNA used in this curve?
    a. virus
    b. bacterium
    c. eukaryote
    d. none of the above

12. Fluorescent *in situ* hybridization is used to:
    a. locate sequences of DNA on a chromosome.
    b. identify mini- and microsatellite DNAs.
    c. identify homologous pairs of chromosomes.
    d. identify the centromere portion of chromosomes.

13. Which of the following statements is true of repeated DNA sequences?
    a. All repeated sequences are coding, but not all coding sequences are repeated.
    b. All nonrepeated sequences are coding, and all coding sequences are nonrepeated.
    c. All repeated sequences lack coding functions, but not all coding sequences are nonrepeating.
    d. Some coding sequences are repeated; others are nonrepeating.

14. Human cells contain several different proteins that bind calcium. They share similar amino acid sequences but are found in different cell types. What can you infer about the genes for these proteins?
    a. They are all part of the family of globin genes.
    b. They probably arose in evolution by gene duplication.
    c. They are part of a family of pseudogenes.
    d. They must all reside on different chromosomes.

15. Cultured hamster cells treated with methotrexate become resistant to the drug by amplifying the gene for DHFR. This is an example of:
    a. selective gene amplification.
    b. random gene duplication and selective survival.
    c. evolution.
    d. all of the above.

16. In eukaryotes, transposable elements arise by:
    a. reverse transcription of RNA, followed by insertion into the genome.
    b. excision of a portion of the chromosome, followed by insertion at a new place in the genome.
    c. duplication of tranposable element's DNA followed by insertion into a new place in the genome.
    d. All of the above.

17. Which of the following is not a role that has been suggested for transposable elements in the process of evolution?
    a. Transposable elements bring previously unlinked segments together, posibly resulting in composite protein products.
    b. Insertion of transposable elements into the genome results in gene duplication and the evolution of multigene families.
    c. Transposable elements have no role in evolution; they are genetic parasites.
    d. Natural selection always favors species with many transposable elements.

18. Which of the following is *not* a reason that the actual number of human genes is still a mystery?
    a. The human genome is nowhere near completion.
    b. It is not certain whether particular sequences of DNA actually contain genes.
    c. Small genes with single exons can be missed by the analytical computers.
    d. The genome's "dark matter" has yet to be accurately sequenced.

19. The difference between genomic DNA and cDNA is
    a. Genomic DNA is that which is expressed and cDNA is all the DNA.
    b. Genomic DNA is found in the nucleus and cDNA is found in the cytoplasm.
    c. Genomic DNA is all the DNA and cDNA is made only from genes that are expressed.
    d. There is no difference between these two types of DNA.

20. Individuals that share a haplotype are most likely to
    a. have all of the same genetic information.
    b. have the same SNPs in the region of the genome that contains that haplotype.
    d. be of different ethnic descent.
    e. be siblings.

Problems and Essays

1. The cells of human embryos undergo division every 24 hours, and each cell has about ten million ribosomes. A cell with one copy of the ribosomal RNA gene and the maximum number of polymerase enzymes transcribing it can produce a ribosome every 1.16 seconds. What is the minimum number of copies of the rRNA genes required to support human embryonic development?

2. The human genome has hundreds of thousands of *Alu* sequences scattered throughout it, and many shorter *Alu*-like sequences that occur between genes, and even within introns. The *Alu* sequence has remarkable similarities to an RNA called 7SL RNA, a molecule that aids in protein secretion. The 7SL RNA is highly conserved in mice, humans, *Drosophila*, and a similar gene has even been identified in *E. coli*, yet the *Alu* SINE is characteristic of mammals. Give a possible explanation for the appearance of the *Alu* sequence.

3. Genome sequence data has proven valuable not only to biomedical researchers, but also to evolutionary biologists interested in the phylogenetic relatedness of different species. Below is a phylogenetic tree showing the relatedness of several primate species. Each branch point represents some set of characteristics that were shared by the common ancestor of the animals at the ends of the branches. The horizontal axis is a rough estimate of the passage of time, with earlier forms on the left and more recent ones on the right. The more recently a branch point occurs, the more closely related those species at the ends of the branches are believed to be. The branch points representing the gorilla, the chimpanzee, and the human are missing.

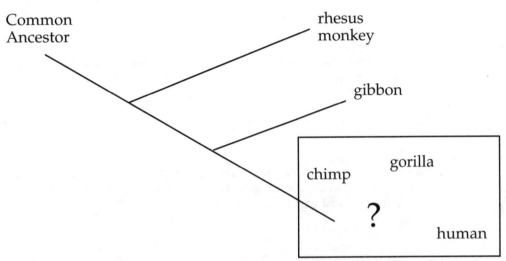

Below are DNA sequences known to encode part of the globin gene from the genomes of three species:

Human: A-G-G-C-A-T-A-A-A-C-C-A-A-C-C-G-A-T-T-A
Chimpanzee: A-G-G-C-C-C-C-T-T-C-C-A-A-C-C-G-A-T-T-A
Gorilla: A-G-G-C-C-C-C-T-T-C-C-A-A-C-C-A-G-G-C-C

a. What do these sequences suggest about the evolutionary relatedness of gorillas, humans and chimps? Draw the missing branch points in the box below, and explain your answer.

b. Suppose you are able to sequence the DNA of this region of the genome of the common ancestor to all three species. You have found this sequence:

Common Ancestor: A-G-G-C-C-G-G-C-T-C-C-A-A-C-C-A-G-G-C-C

Does this change your answer to part a above? Why or why not?

4. The human genome has three billion base pairs on 23 chromosomes, and an estimated 30,000 genes.

    a. What is the average number of genes on each chromosome?

    b. What is the average number of base pairs for each gene?

    c. If 95% of the base pairs are noncoding DNA, what is the average number of base pairs for each gene?

5. Draw the reannealing curve that might result for DNA extracted from yeast cells. Which part of the curve represents the DNA from the centromere portion of the yeast chromosomes? Which part of the curve represents the actual yeast genes?

6. Two genes, one for seed color and one for seed shape, were studied in dandelion plants and found to obey Mendel's law of independent assortment. When the genes were located by in situ hybridization, however, they were found to reside on the same chromosome. What explanation can you provide for this observation? In the diagram, show where the fluorescence of biotinylated DNA probes might be concentrated for these two genes.

7. You are searching for the gene for a hereditary disease using differences in the lengths of DNA fragments that result from digestion of different people's DNA with a specific restriction endonuclease. (Such diffrerences are called restriction fragment length polymorphisms.) You have isolated DNA from four people with the disease, and four people without the disease, and treated the DNA with a restriction endonuclease. Agarose gel electrophoresis resulted in the banding patterns below. Which band probably represents the fragment of DNA closest to the gene for the disease? Why?

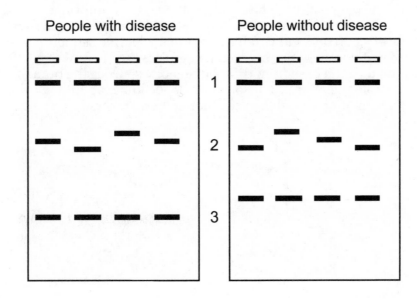

Web Explorations

Anyone with an internet connection can access the human genome, at least the one produced by the International Human Genome Sequencing Consortium. Why not give it a try?

Begin by going to the National Center for Biotechnology Information web site at http://www.ncbi.nlm.nih.gov/. Find the entry into the human genome resources and click on one of the chromosomes, for example the Y chromosome. You will see a list of genes mapped to the Y chromosome. Now choose a gene that sounds interesting, say the sex determining factor gene. Click on the name of the gene of interest to go to an information screen about it. Among the information on that screen will be a series of identification numbers, called Accession numbers. Click on one and you may get the nucleotide sequence or the amino acid sequence of the protein it encodes. See if you can negotiate the NCBI web site to learn how to compare sequences on the data base from different genomes. Does the human sex determining factor gene have a homologue in yeast? If so, what do we know about the protein in yeast? How is this kind of information valuable to cell biology researchers?

11 Expression of Genetic Information: From Transcription to Translation

Learning Objectives

When you have finished this chapter, you should be able to:

1. Understand the concept of information flow in the cell.
2. Define transcription and summarize the main differences between eukaryotic and prokaryotic transcription.
3. Know the roles of each of the three RNA polymerases in eukaryotes.
4. Understand the concept of RNA processing and modification.
   a. Describe the mechanism of rRNA processing in eukaryotes.
   b. Define split gene, intron, exon, and RNA splicing.
   c. Know the importance of mRNA modifications on 5' and 3' ends.
5. Know the main features of the genetic code.
6. Explain the role of tRNA in decoding the genetic information.
7. Give an overview of translation and specify the factors involved in each step.

Key Terms and Phrases

gene	transcription	translation
RNA polymerase	promoter	processivity
core enzyme	sigma factor (σ)	Pribnow box
rho factor	RNA polymerase I	RNA polymerase II
RNA polymerase III	transcription factors	primary transcript (pre-RNA)
small nuclear RNA (snRNA)	transcription unit	ribosomal RNA (rRNA)
nucleolus	nucleolar organizer	Svedberg unit (S value)transfer
RNA (tRNA)	messenger RNA (mRNA)	core promoter element
heterogeneous nuclear RNA	(hnRNA)	snoRNA
preinitiation complex	TATA-binding protein (TBP)	TATA box
methylguanosine cap	poly(A) tail	split genes
intervening sequences	exon	intron
RNA splicing	splice sites	spliceosome
snRNPs (snurps)	group I intron	group II intron
alternate splicing	exon shuffling	ribozyme
genetic code	codon	reading frame
frameshift mutation	degenerate code	anticodon
wobble hypothesis	aminoacyl-tRNA synthetase	mutagen
translation	initiation	initiation codon
Shine-Dalgarno sequence	initiation factor (IF)	A (aminoacyl) site
P (peptidyl) site	elongation	elongation factors (EF)
peptidyl transferase	translocation	termination
release factors	nonsense mutation	polyribosome (polysome)

Reviewing the Chapter

I. The Relationship between Genes and Proteins
    A. Genes store information for producing all cellular proteins.

    B. Early observations suggested a direct relationship between genes and proteins.
        1. The relationship between a specific gene, a specific enzyme and a specific metabolic condition was first described by Archibald Garrod in 1908 in studies of "inborn errors of metabolism."
        2. The "one gene–one enzyme" hypothesis was formulated by Beadle and Tatum. Later it was modified to "one gene–one polypeptide chain."
        3. Mutation in a single gene causes a single substitution in an amino acid sequence of a single protein.

    C. An overview of the flow of information through the cell:
        1. Messenger RNA is an intermediate between a gene and its polypeptide.
        2. Transcription is the process whereby RNA is formed from a DNA template.
        3. Proteins are synthesized in the cytoplasm by the process of translation from an mRNA template.
        4. There are three classes of RNA in a cell: messenger RNA, ribosomal RNA, and transfer RNA.
        5. Activity of both rRNA and tRNA depends on their folding into well-defined secondary and tertiary structures.
        6. RNAs often contain mismatched base pairs and modified nitrogenous bases (unlike DNA).

II. An Overview of Transcription and Translation in Both Prokaryotic and Eukaryotic Cells
    A. DNA-dependent RNA polymerase assembles a linear chain of nucleotides which is complementary to the template DNA strand.
        1. The promoter is the site at the 5′ end of the gene, where RNA polymerase binds prior to initiating transcription.
        2. The sequence at which the RNA polymerase initially binds to the DNA is called the promoter.
        3. The newly synthesized RNA chain grows in a 5′ to 3′ direction, antiparallel to the DNA.
            a. RNA polymerase must be processive– remain attached to DNA over long stretches of template.
            b. RNA polymerase must be able to move from nucleotide to nucleotide.
        4. Nucleotides enter the polymerization reaction as trinucleotide precursors. The reaction is driven forward by the exergonic hydrolysis of a pyrophosphate.
        5. After the polymerase has passed, the DNA-RNA hybrid dissociates and the DNA double helix reforms.
        6. The enzymatic activities of RNA polymerase, digestion and polymerization, are catalyzed by the same active site.

    B. Transcription in prokaryotes:
        1. There is only one type of RNA polymerase in prokaryotic cells.
        2. Transcription-competent prokaryotic RNA polymerase consists of a core enzyme and a sigma factor. The sigma factor dissociates after transcription has begun.
        3. Bacterial promoters are located upstream from the site of initiation.
            a. They have two conserved regions: the -35 region and the Pribnow box.
            b. The most common version of the conserved sequence is called the consensus sequence.

4. Differences in the DNA sequences at both -35 and the Pribnow box may regulate gene expression.
5. Termination in prokaryotes can either require a rho factor or be rho-independent.

C. Transcription and RNA Processing in Eukaryotic Cells
1. There are three types of RNA polymerases in eukaryotes:
2. Different classes of RNAs are synthesized by different RNA polymerases.
   a. Most ribosomal RNAs are transcribed by RNA polymerase I.
   b. Messenger RNAs are transcribed by RNA polymerase II.
   c. Transfer RNAs are transcribed by RNA polymerase III.
3. General and specific transcription factors regulate the activity of RNA polymerases.
4. The newly transcribed RNAs are processed.
   a. A primary transcript is the initial RNA molecule synthesized.
   b. A transcription unit is the segment of DNA corresponding to a primary transcript.
   c. Small nuclear RNAs (snRNAs) and their associated proteins are required for RNA processing.

III. Synthesis and Processing of Ribosomal and Transfer RNAs

A. Ribosomal RNAs:
1. Ribosomes are produced in nucleoli.
   a. Nucleolar organizers are regions of a chromosome that contain rRNA genes.
   b. The fibrillar core of the nucleolus consists of rDNA templates and nascent rRNA transcripts.
   c. Both ribosomal subunits are assembled in the nucleolus.
2. Synthesizing the rRNA precursor:
   a. Ribosomal RNA genes are arranged in tandem.
   b. Ribosomal RNA transcription has a "Christmas tree" pattern.
   c. Proteins that convert the rRNA precursor into mature rRNA become associated with pre-rRNA during transcription.
   d. Nontranscribed spacers separate various transcription units in a gene cluster.
3. Processing the rRNA precursor:
   a. By splicing, 28S, 18S and 5.8S rRNAs are derived from a single 45S primary transcript.
   b. Pre-rRNA contains large numbers of methylated nucleotides and pseudouridine residues.
   c. Unaltered sections of the pre-RNA are discarded.
   d. Small, nucleolar RNAs (snoRNAs) help with pre-RNA processing.
      (1) snoRNAs are packaged with proteins into small, nucleolar ribonucleoproteins (snoRNPs).
      (2) snoRNAs modify RNA bases in pre-rRNAs.
   e. rRNA and protein are assembled into ribosomes in the nucleolus.
4. Synthesis and processing of the 5S rRNA:
   a. The 5S rRNA genes are located outside the nucleolus.
   b. A 5S rRNA is transcribed by RNA polymerase III.
   c. RNA polymerase III uses an internal promoter.

B. Transfer RNAs:
1. Transfer RNA genes are located in small clusters scattered around the genome.
2. Transfer RNAs have promoter sequences within the coding region of the gene; the promoter is split into two portions.

3. During processing, tRNA precursor is trimmed and the CCA triplet is added to its 3′ end.

IV. Synthesis and Processing of Messenger RNAs:
A. Heterogeneous nuclear RNAs are precursors of mature mRNAs.
   1. Heterogeneous nuclear RNAs (hnRNAs) are found only in nucleus.
   2. hnRNAs have large molecular weights.
   3. hnRNAs are very unstable and have half-lives of a few minutes.
B. The machinery for mRNA transcription:
   1. RNA polymerase is assisted by general transcription factors (GTFs), whose precise role is unknown.
   2. The core promoter element for polymerase II lies 24 to 32 bases upstream from the initiation site; this region contains the TATA box.
   3. The preinitiation complex of general transcription factors and polymerases assemble at the TATA box.
   4. The preinitiation complex assembly starts with the binding of the TATA-binding protein (TBP) to the promoter.
   5. TBP is a subunit of the protein TFIID and when it binds to the promoter it causes a dramatic distortion in the conformation of the DNA.
   6. RNA polymerase is heavily phosphorylated in the CTD region. TFIIH acts as the protein kinase.
C. Messenger RNAs share certain properties:
   1. They each code for a specific polypeptide.
   2. They are found in the cytoplasm.
   3. They are attached to ribosomes when translated.
   4. They have both noncoding portions and modifications at their 5′ (guanosine cap) and 3′ (poly A tail) ends.
D. Split genes– An unexpected finding:
   1. Eukaryotic genes contain intervening sequences called introns. These sequences are missing from mature mRNAs.
   2. Exons are the parts of a split gene which are included in the mature mRNA.
   3. Hybridization experiments supported the concept of a split gene. Loops in the hybrid RNA-DNA were the introns.
E. The processing of eukaryotic messenger RNAs:
   1. During processing, 5′ methylguanosine caps and 3′ poly(A) tails are added to mRNAs.
   2. Intervening sequences are removed and exons are connected by RNA splicing.
      a. Sequences between exon-intron junctions are highly conserved.
      b. Spliceosomes facilitate mRNA splicing.
      c. Spliceosomes consist of snRNPs (snRNAs bound to proteins).
   3. Group I and group II introns are capable of self-splicing.
   4. Introns are removed in a preferred order.
F. Evolutionary implications of split genes and RNA splicing:
   1. The theory of "an RNA world" suggests that RNA was the earliest molecule to both store information and catalyze reactions.
   2. RNA splicing via spliceosomes is thought to have evolved from self-splicing RNAs.
   3. Exon shuffling could have played a major role in the evolution of many genes.
G. Creating new ribozymes in the laboratory:
   1. Ribozymes are RNAs that have catalytic activity.
   2. Modified RNAs can catalyze certain reactions.

3. Synthesis of random RNA molecules is one of the strategies used to look for ribozymes.

V. Small Noncoding RNAs and RNA Interference
A. RNA interference (RNAi) results in the destruction of certain mRNAs.
    1. Double-stranded RNAs are taken up by cells where mRNAs with the same sequences are then destroyed.
    2. The dsRNA is cleaved into small-interfering RNAs (siRNA) by Dicer.
    3. The siRNA fragments are incorporated into a protein complex, called RISC.
    4. RISC guides the siRNA to complementary mRNAs.
    5. An enzyme of RISC complex destroys the complementary mRNAs.
    6. Mammalian cells respond to very small dsRNAs of 20 or 21 base pairs.
B. MicroRNAs: Hundreds of RNAs of unknown function.
    1. Very small RNAs of 20 to 21 base pairs are called microRNAs (miRNAs).
    2. miRNAs play a role in development, gene regulation, and genomic rearrangements.

VI. The Human Perspective: Potential Clinical Applications of RNA Interference
A. Strategies for using RNAi to combat disease are being developed against cancer, viruses, and some genetic disorders.
B. siRNAs directed against mRNAs characteristic of disease block synthesis of abnormal proteins.
C. Although the technique is promising, some obstacles remain:
    1. Viral genes mutate quickly, making former siRNAs ineffective.
    2. Delivering genes for siRNAs can lead to complications.

VII. Encoding Genetic Information
A. Information stored in a gene is present in the form of a genetic code.
B. The properties of the genetic code:
    1. The genetic code uses triplet, nonoverlapping, and degenerate codons for each amino acid.
    2. Codon assignment was determined by transcription from artificial mRNAs.
    3. Codon assignment is nonrandom; codons for the same amino acid are similar.
    4. The first two codon bases for a particular amino acid are invariant, whereas the third base may vary.

VIII. Decoding the Codons: The Role of Transfer RNAs
A. Transfer RNAs are a class of adapter molecules that recognize individual amino acids and the corresponding mRNA codons.
    1. The structure of transfer RNAs
        a. The amino acid is attached to the 3' end of tRNA.
        b. The secondary structure of tRNA resembles a clover leaf, while the tertiary structure is an L shape.
        c. The anticodon on tRNA complements the codon of the mRNA.
        d. The wobble hypothesis suggests that a single kind of tRNA can recognize codons with variable third bases.
    2. Amino acid activation:
        a. Specific aminoacyl-tRNA synthetases link amino acids with their respective tRNAs.
        b. Energy of one ATP is used to activate the amino acid, which is then transferred to the tRNA molecule.
        c. Codons of the mRNA are ultimately interpreted according to the recognition abilities of the aminoacyl-tRNA synthetases.

IX. Translating Genetic Information
A. Translation is the most complex synthetic activity of the cell.

B. Initiation:
    1. Translation begins at the initiation codon, AUG, in the mRNA which automatically puts the ribosome in the correct reading frame.
    2. The small ribosomal subunit identifies the correct AUG codon.
        a. In bacterial mRNAs, the Shine-Dalgarno sequence guides the subunit to the correct initiation codon.
        b. In eukaryotes, the smallest ribosomal subunit recognizes the 5′ end of the message and finds the first AUG triplet by scanning.
        c. Initiation factors are required for initiation.
    3. AUG codes for methionine so it is always the first amino acid to be incorporated into a polypeptide chain.
    4. There are two different methionyl-tRNAs: one for protein synthesis and one for the methionyl residues in the polypeptide.
    5. After the initiator tRNA is bound the large subunit joins the complex.
    6. Ribosomes have two sites for tRNA molecules: the A (aminoacyl) site and the P (peptidyl) site.

[handwritten in left margin: identification of AUG methionine]

C. Elongation:
    1. The elongation cycle is the process of the orderly addition of each subsequent amino acid to the growing polypeptide chain.
        a. With the charged amino acid in the P site, the next aminoacyl-tRNA binds to the vacant A site.
        b. Several elongation factors (EFs) are required for elongation.
    2. Peptidyl transferase, one of the rRNAs of the large subunit, forms a peptide bond between amino acids.
    3. In translocation, the ribosome moves one codon in the 3′ direction transferring the deacylated tRNA from the A site to the P site.
    4. At least two molecules of GTP are hydrolyzed during elongation.
    5. Mutations resulting in an addition or deletion of an amino acid are called frame-shift mutations and result in the production of an entirely abnormal sequence of amino acids from the point of the mutation.
    6. Messenger RNA contains recoding signals that cause ribosomes to change reading frame.

D. Termination:
    1. Termination occurs at one of three stop codons: UAA, UAG, or UGA.
    2. Release factors interact directly with the stop codons.
    3. Nonsense mutations produce stop codons and cause premature chain termination.

E. mRNA surveillance: No nonsense allowed:
    1. Nonsense mutations that result in inappropriate stop codons are destroyed.
    2. The process is called nonsense-mediated decay (NMD).
    3. Spliced mRNA contains proteins called the exon-junction complex (EJC).
    4. EJCs on untranslated RNAs are a signal that an inappropriate stop has arisen.

F. Polyribosomes are complexes of multiple ribosomes on mRNA, allowing simultaneous translation, which increases the rate of protein synthesis.

X. Experimental Pathways: The Role of RNA as a Catalyst

A. RNA is capable of catalyzing a complex, multistep reaction of self-splicing.
B. The RNA portion of ribonuclease P can accurately cleave the tRNA precursor.
C. The RNA component plays an important role in the activity of ribosomal peptidyl transferase.

Key Figure

Figure 11.2. An overview of the flow of information in a eukaryotic cell.

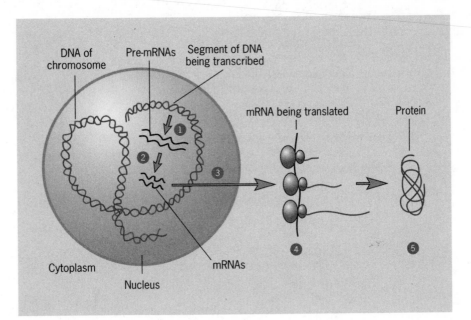

Questions for Thought:

1. How would a diagram depicting the flow of information in a prokaryotic cell differ from this one?
2. In most cases, RNA acts as an intermediary molecule between the genetic information and the final product of that information, proteins. What are the advantages of having an intermediary as opposed to translating the DNA directly into protein?
3. What is the metabolic cost (in either ATP or GTP) of each of these steps? Which step is most "expensive" to the cell?
4. Where in this process are errors likely to occur, and what mechanisms exist to minimize errors?

Review Problems

Short Answer

1. Give the significance (not the definition) of the following terms or phrases. Say what they do or why they are important. For example:
    translation: *interprets the sequence of nucleotides in mRNA into a sequence of amino acids in proteins.*

    a. mRNA as an intermediary between DNA and protein

    b. the highly exergonic nature of RNA polymerization

    c. slight differences in the sequences of nucleotides in promoter regions

    d. methylguanosine cap

    e. alternate splicing

    f. general transcription factors (as a group)

    g. Shine-Dalgarno sequence

    h. our changing concept of the gene

2. Compare and contrast the following:

a. "one gene– one enzyme" vs. "one gene– one transcription unit"

b. higher-order structure of proteins vs. higher-order structure of RNA

c. Pribnow box vs. TATA box

d. upstream promoters vs. internal promoters

e. frameshift mutations vs. substitution mutations

Multiple Choice

1. Why was *Neurospora* a good organism for Beadle and Tatum's studies?
   a. *Neurospora* grows well in rich media and cannot synthesize many enzymes.
   b. *Neurospora* grows well in minimal media, and is sensitive to enzymatic deficiencies.
   c. *Neurospora* requires many vitamin additives to grow.
   d. *Neurospora* was actually a poor choice.

2. The tertiary structure of RNA is mostly dependent upon:
   a. the tendency to bury hydrophobic nucleotides within the core of the molecule.
   b. the formation of additional covalent bonds between nonadjacent nucleotides.
   c. the formation of complementary base pairs between nonadjacent nucleotides.
   d. the formation of salt bridges between the sugar-phosphate backbones.

3. When used to describe the RNA polymerases, the term "processive" refers to:
   a. the ability to recognize and bind to a promoter region.
   b. the ability to remain attached to DNA and form long RNAs.
   c. the ability to add nucleotide triphosphates to a growing RNA chain.
   d. all of the above.

4. Which of the following lists of transcription features applies only to prokaryotes?
   a. Pribnow box, rho factor, core enzyme, -35 region
   b. TATA box, primary transcript, RNA polymerase III, transcription factors
   c. Pribnow box, transcriptional unit, rho factor, RNA polymerase III
   d. TATA box, rho factor, core enzyme, -35 region

5. Of the following statements about the different types of eukaryotic RNA, which one is true?
   a. Over 80% of cellular RNA is ribosomal, yet most newly synthesized RNA is mRNA.
   b. The genes for rRNA and tRNA occur in clusters, whereas most genes for mRNA do not.
   c. All three types of RNAs are processed before reaching their mature form.
   d. All of the above statements are true.

6. The significance of methylation of pre-rRNA is probably:
   a. to ensure that the final rRNA will assume the correct tertiary structure.
   b. to prevent the pre-rRNA from being processed.
   c. to prevent degradation of the parts destined to become mature rRNA .
   d. none of the above.

7. If cultured cells are incubated with ^{14}C-methionine, the radiolabel will appear:
   a. in all the rRNAs simultaneously.
   b. first in the 45S RNA, followed by the 32S RNA, followed by the 28S RNA.
   c. first in the 45S RNA, followed by the 28S RNA, followed by the 5S RNA.
   d. first in the 28S RNA, followed by the 32S RNA, followed by the 45S RNA.

8. You have examined nucleoli structures from several sources listed below and find that the cell type with the largest nucleolus is the:
   a. human nerve cell.
   b. frog muscle cell.
   c. frog oocyte.
   d. rat liver cell.

9. The 5S rRNA differs from the other cellular rRNAs in which of the following ways?
   a. The genes are located outside the nucleolus.
   b. The genes are transcribed by RNA polymerase III.
   c. The promoter region of the 5S rRNA gene is internal to the gene.
   d. All of the above.

10. Which of the following statements is true of tRNAs?
    a. There are the same number of tRNA genes as there are different tRNAs.
    b. There are more tRNA genes than different tRNAs, and more tRNAs than amino acids.
    c. There are more different tRNAs than there are different triplet codons.
    d. There are the same number of tRNAs as there are amino acids.

11. In a cell that produces only one type of protein, which of the following statements about the relationship between hnRNA and mRNA is true?
    a. Heterogeneous nuclear RNA is always larger than mRNA.
    b. Heterogeneous nuclear RNA is always smaller than mRNA.
    c. Heterogeneous nuclear RNA and mRNA are typically the same size.
    d. There is no consistent relationship between the size of the two molecules.

12. One way to tell if RNA polymerase II is in the process of transcribing an mRNA is to:
    a. check for the presence of a repeated 7-amino acid sequence at the C-terminal domain.
    b. check for phosphorylation at the C-terminal domain.
    c. both a and b
    d. neither a nor b

13. Using both the enzyme reverse transcriptase and radiolabeled deoxynucleotides, you make a radioactive DNA copy from an mRNA. When you hybridize that copy DNA to nuclear DNA, you find that there are seven different chromosomal positions where hybridization occurs, and there are three unhybridized loops at each of the seven positions. From this you conclude:
    a. There are seven copies of the gene for that mRNA and three nontranscribed segments in each gene.
    b. There are seven copies of the gene for that mRNA and three exons in each gene.
    c. There are three copies of the gene for that mRNA and seven exons in each gene.
    d. There are seven copies of the gene for that mRNA and three introns in each gene.

14. Why is the methylguanosine cap added to mRNA before the poly(A) tail?
    a. because the cap is added to the 5′ end before the 3′ end is fully transcribed
    b. because the cap is added to the 3′ end before the 5′ end is fully transcribed
    c. because the cap is required for the addition of the poly(A) tail
    d. none of the above

15. The textbook states that the nucleotide sequence of splice sites is "highly conserved" and of "ancient origin." How do we know this sequence is ancient?
    a. It is widespread and occurs in even the most primitive eukaryotes.
    b. The number of mutations that have accumulated in the sequence indicates that it must have been in cells for millions of years.
    c. It is only found in organisms that are highly evolved.
    d. It codes for ancient proteins.

16. About 30% of point mutations (substitutions) that occur in coding regions of DNA are harmless. Why?
    a. because only 30% of the DNA is actually coding for amino acids
    b. because only 70% of the DNA is actually coding for amino acids
    c. because most amino acids are specified by several codons that differ at the third position
    d. all of the above

17. Which type of mutation is likely to be the most harmful?
    a. a base substitution in which a valine is replaced with a leucine.
    b. a +3 frameshift mutation.
    c. a base substitution in which a glutamic acid is replaced with an aspartic acid.
    d. a +1 frameshift mutation.

18. Of the following statements about the aminoacyl tRNA synthetases, which one must be true?
    a. There are at least as many aminoacyl tRNA synthetases as there are amino acids.
    b. There are at least as many aminoacyl tRNA synthetases as there are tRNAs.
    c. There are at least as many aminoacyl tRNA synthetases as there are mRNAs.
    d. There are at least as many aminoacyl tRNA synthetases as there are rRNAs.

19. The following complex of macromolecules is found in a cell: DNA attached to RNA polymerase, attached to a growing chain of RNA, attached to a ribosome, attached to two tRNAs, attached to a growing chain of amino acids. The cell must be:
    a. a plant cell.
    b. an animal cell.
    c. a virus.
    d. a prokaryotic cell.

20. Which of the following observations supports the view that ribosomal RNA has catalytic activity?
    a. The base sequence of rRNAs is highly conserved, whereas the amino acid sequence of ribosomal protein is not.
    b. Bacteria that develop antibiotic resistance have base substitutions on the rRNA, but do not have amino acid substitutions on the ribosomal proteins.
    c. The peptidyl transferase reaction is sensitive to ribonucleases.
    d. All of the above.

Problems and Essays

1. Eukaryotic RNA polymerases recognize a promoter that most often has the sequence TATAA. On which strand of DNA, the actual template strand or its complementary partner, does the TATA box sequence occur? Explain.

2. Indicate where in a eukaryotic cell you would be most likely to find each of the following types of RNA.

    _____hnRNA   _____tRNA

    _____mRNA   _____pre-RNA

    _____snRNA   _____nascent rRNA

    _____rRNA   _____45S pre-rRNA

3. A certain gene is 10 kilobases in length, and has two introns and three exons. Within the gene are two restriction sites for EcoR1, as shown. When the mature mRNA is reverse transcribed, and the cDNA is treated with EcoR1, how many restriction fragments result, and what is the length of each?

4. Niremberg and Matthaei used the enzyme polynucleotide phosphorylase to assemble RNAs with random sequences in the absence of a template. Suppose they added the enzyme to a solution of 90% CTP and 10% UTP, and then used a cell-free system of protein synthesis to make peptides according to the sequences in their synthetic RNAs.

    a. Which amino acids could possibly appear in the synthetic protein?

    b. Which amino acid would be most abundant in the synthetic protein?

    c. Which amino acid would be least abundant?

5. Starting with a DNA template and the appropriate machinery for both transcription and translation, how many nucleotide triphosphates are needed to synthesize an octapeptide? Why is the recommended caloric intake (per pound) of a growing child higher than that for an adult?

6. A bacterium produces a normal protein with the following amino acid sequence:

    MET-VAL-HIS-LYS-ARG-THR-LEU-VAL

    After irradiation, a mutant strain is produced that makes a mutant protein from the same coding site on DNA:

    MET-VAL-HIS-LYS-GLU-PRO

    Describe the mutation that occurred.

7. Starting with a solution of randomly-generated synthetic RNAs, design an experiment to select for those RNAs that can bind to ATP.

Web Explorations

Until recently, the streptococcus bacterium that causes some pneumonias , meningitis and ear infections was rather easily treated using antibiotics. In the last 10 years, however, several antibiotic resistant strains have appeared. In 1991, only 1.4 percent of streptococcal bacterial infections were antibiotic resistant. By 2000, that number had risen to 14 percent. Not surprisingly, scientists have been anxious to uncover the molecular mechanisms of antibiotic resistance with the hope of developing better treatments for these resistant pathogens.

To date, several molecular mechanisms of antibiotic resistance have been identified. One study found that 44 percent of resistant bacterial strains underwent ribosomal modification by a methylase specified by the ermB gene. Using the World Wide Web, learn how the ermB gene confers antibiotic resistance to streptococcus. Begin by learning how antibiotics—especially penicillin-based antibiotics-- affect bacterial translation. Then investigate the manner in which ribosomal modification foils the action of antibiotics. Can you propose a mechanism for overcoming the ermB gene modification?

12

The Cell Nucleus and the Control of Gene Expression

Learning Objectives

When you have finished this chapter, you should be able to:

1. Describe the eukaryotic nucleus, including the morphological features of mitotic chromosomes and nonmitotic chromatin.
2. Know the structure and function of nuclear pores. List the substances that move between the nucleus and the cytoplasm, and the means by which they move.
3. Understand the role of histone proteins in chromatin packaging.
4. Understand the concept of an operon. Give examples of inducible and repressible operons.
5. Explain the concept of differentiated cells in a eukaryotic organism.
6. Define the levels on which regulation of gene expression can take place. Know the role of transcription factors, methylation, DNA packaging, alternative splicing, and mRNA stability in the regulation of gene expression, and give examples of each type of regulation.

Key Terms and Phrases

chromosomes
nucleolus
perinuclear space
nuclear matrix
histones
nucleosome core particle
mitotic chromosomes
euchromatin
position effect
genetic mosaics
centromere
chromosome aberrations
deletion
gene induction
negative control
promoter
inducer
catabolite gene activator protein (CAP)
inducible operon
selective gene amplification
general transcription factors
basal level elements
palindrome
helix-loop-helix (HLH)
processing-level control
untranslated regions (UTRs)
masked mRNA

chromatin
nucleoplasm
nuclear lamina
nuclear pore complex
nonhistone chromosomal proteins
linker DNA
lampbrush chromosomes
constitutive heterochromatin
Barr body
karyotype
telomere
inversions
cri du chat syndrome
gene repression
operon
operator
gene regulatory protein
trp operon
transcriptional-level control
translational-level control
glucocorticoid response elements
transcription factor motifs
genomic imprinting
high mobility group (HMG) proteins
DNase I hypersensitive sites
poly (A) binding protein

nuclear matrix
nuclear envelope
lamins
nuclear localization signal
nucleosome
packing ratio of DNA
heterochromatin
facultative heterochromatin
X inactivation
Q bands
telomerase
translocation
duplication
positive control
structural genes
repressor
regulatory gene
repressible operon
lac operon
transcription factors
metal response elements
glucocorticoid receptor
zinc finger
DNA methylation
alternate splicing
enhancers
translational frameshifting

mRNA editing   translational bypassing   protein splicing
homeobox   homeodomain   Hox genes
N-end rule   ubiquitin

Reviewing the Chapter

I. Introduction
    A. All cells in a multicellular organism contain the same complement of genes.
    B. Cells express their genetic information selectively.
    C. Gene expression is controlled by regulatory machinery in the cell nucleus.

II. The Nucleus of a Eukaryotic Cell
    A. The contents of the nucleus are enclosed by a nuclear envelope. A typical nonmitotic nucleus includes several major components:
        1. Chromosomes are present as extended fibers of chromatin.
        2. The nuclear lamina is a protein-containing fibrillar network.
        3. Nucleoli are the sites of ribosomal RNA synthesis.
        4. Nucleoplasm is the fluid in which the solutes of the nucleus are dissolved.
    B. The nuclear envelope:
        1. The nuclear envelope is a complex structure at the boundary between the nucleus and the cytoplasm of a eukaryotic cell.
            a. A nuclear envelope consists of two membranes separated by a perinuclear space.
            b. The two membranes are fused at sites forming a circular nuclear pore.
            c. The inner surface of the nuclear envelope is lined by the nuclear lamina.
                (1) The nuclear lamina supports the nuclear envelope.
                (2) The nuclear lamina is composed of lamins.
                (3) The integrity of the nuclear lamina is regulated by phosphorylation and dephosphorylation.
        2. The structure and function of the nuclear pore complex:
            a. Proteins and RNAs are transported in and out of the nucleus.
            b. Bidirectional transport occurs through the nuclear pore complex (NPC).
            c. Proteins synthesized in the cytoplasm are targeted to the nucleus by the nuclear localization signal (NLS).
                (1) An NLS usually consists of one or two short stretches of positively-charged amino acids.
                (2) Proteins with an NLS stretch bind to an NLS receptor.
                (3) The NLS receptor docks proteins to the nuclear pore complex.
                (4) Conformation of the NPC changes as the protein passes through.
                (5) RNAs move through the NPCs as RNPs (ribonucleoproteins) and carry NES (nuclear export signals) sequences that allow them to pass through the envelope.
    C. Chromosomes and Chromatin:
        1. Chromosomes are visible only during mitosis.
        2. Packaging the genome:
            a. Chromosomes consist of chromatin fibers, composed of DNA and associated protein.
                (1) Each chromosome contains a single, continuous DNA molecule, packaged to occupy a small space.

(2) The protein component of chromosomes includes histones and nonhistone chromosomal proteins.

  b. Nucleosomes: The lowest level of chromosome organization:

    (1) Histones are a group of very basic proteins that are involved in DNA packaging.

      (a) Histones are highly conserved between different species.

      (b) DNA and histones are organized into repeating subunits called nucleosomes.

      (c) Each nucleosome includes a core particle and a linker.

      (d) DNA is wrapped around the core complex.

      (e) The histone core complex consists of two molecules each of histones H2A, H2B, H3, and H4 forming an octamer.

      (f) H1 histone is located outside of the nucleosome core particle.

    (2) Positively charged residues on histones interact with negatively charged phosphate groups of the DNA backbone.

      (a) Positively charged residues occur in a spiral along the path of DNA wound around the core.

      (b) DNA-histone interaction is not sequence specific.

    (3) Histone genes are present in the genome as tandemly repeated sequences.

    (4) Binding of a nonhistone protein to DNA can influence the positioning of nucleosome particles in the adjacent region of the chromosome.

  c. Higher levels of chromatin structure:

    (1) A 30 nm filament is the second level of chromatin structure.

    (2) A 30 nm filament packaging is maintained by histone H1.

    (3) Chromatin filaments are organized into large supercoiled loops.

    (4) The presence of loops in chromatin can be seen:

      (a) in mitotic chromosomes from which histones have been extracted.

      (b) in meiotic lampbrush chromosomes of amphibian oocytes.

3. Heterochromatin and euchromatin:

  a. Euchromatin returns to a dispersed state after mitosis.

  b. Heterochromatin is condensed during interphase.

  c. Heterochromatin can be either constitutive or facultative.

    (1) Constitutive heterochromatin remains condensed all the time.

    (2) Constitutive heterochromatin is mostly situated around centromeres and telomeres.

    (3) Constitutive heterochromatin consists of highly repeated sequences and is devoid of protein-coding genes.

    (4) Active genes placed next to constitutive heterochromatin become silenced as a result of their position.

  d. One of the X chromosomes in females is present as facultative heterochromatin and is called a Barr body.

  e. X inactivation:

    (1) Heterochromatization of one X chromosome in each cell of a female mammal occurs at an early stage in development.

(2) X inactivation is a random process (either of the two X chromosomes can be inactivated), making adult females genetic mosaics.

(3) X inactivation is initiated by an RNA molecule.

4. The histone code and formation of heterochromatin:
    a. The histone code hypothesis states that the activity of a chromatin region depends on the degree of chemical modification of histone tails.
    b. Heterochromatin has many methylated H3 histones, which stabilize the compact nature of the chromatin.
    c. Small RNAs and specific enzymes play a role in histone methylation.

5. The structure of the mitotic chromosome:
    a. Chromatin of a mitotic cell exists in its most highly condensed state.
        (1) The shape of the mitotic chromosome depends on the length of the DNA molecule and the position of the centromere.
        (2) Staining mitotic chromosomes can provide useful information.
        (3) A karyotype is a portrayal of pairs of homologous chromosomes ordered according to decreasing size.
            (a) The pattern of Q bands is highly characteristic for each chromosome of a species.
            (b) Specific genes can be located on mitotic chromosomes by in situ hybridization.
    b. Centromeres:
        (1) The centromere is located at the site of marked indentation on a chromosome.
        (2) Centromeres contain constitutive heterochromatin.
        (3) Centromeric DNA is the site of microtubule attachment during mitosis.
        (4) The site of the centromere does not depend on DNA sequence.
            (a) Sequence-independent characteristics can be inherited.
            (b) Epigenetic inheritance occurs at least partially independently of DNA sequences.
    c. Telomeres:
        (1) The ends of DNA molecules are marked with a set of repeated sequences called telomeres.
        (2) New repeated units are added by a telomerase, which is a reverse transcriptase that assembles DNA from an RNA template.
        (3) Telomeres are required for the complete replication of the chromosome because they protect the ends from being degraded.
        (4) Teleomerase activity is thought to have major effects on cell life.

6. The Human Perspective: Chromosome aberrations:
    a. A chromosomal aberration is a loss or exchange of a segment between different chromosomes, caused by exposure to DNA-damaging agents.
    b. Chromosomal aberrations have different consequences depending on whether they are in somatic or germ cells.
    c. Inversions, translocations, deletions, and duplications are examples of different chromosomal aberrations.
        (1) Inversion involves the breakage of a chromosome and resealing of the segment in a reverse order.
        (2) Translocation is attachment of all or a piece of one chromosome to another chromosome.
        (3) Deletion is the loss of a portion of a chromosome.

(4) Duplication occurs when a portion of a chromosome is repeated.
d. Many human diseases are caused by chromosomal aberrations.
D. Epigenics: There's more to inheritance than DNA sequence
1. Epigenetic inheritance depends on factors other than DNA sequences
2. Histones are distributed randomly to daughter cells during cell division.
a. About half of the parental histones are found in each daughter cell.
b. Parental histones determine the chemical modifications found on the other half of the histones—the newly synthesized histones.
3. DNA that is silent in parental cells is also silent in both daugher cells.
E. The nucleus as an organized organelle:
1. Chromatin fibers are concentrated at specific domains within the nucleus.
2. Chromosome ordering is directed by the nuclear envelope proteins.
3. In the nucleus, mRNAs are synthesized at discrete sites.
4. The nuclear matrix:
a. The nuclear matrix is a network of protein-containing fibrils.
b. The nuclear matrix serves as a skeleton, maintaining the shape of the nucleus and anchoring the machinery involved in nuclear activities.
c. Newly transcribed pre-mRNAs are held in place by elements of the nuclear matrix.

III. **Control of Gene Expression in Prokaryotes**
A. Bacterial cells selectively express genes to use the available resources efficiently.
1. The presence of lactose in the medium induces the synthesis of ß-galactosidase.
2. The presence of tryptophan in the medium represses the genes that encode enzymes for tryptophan synthesis.
B. The bacterial operon:
1. An operon is a functional complex of genes containing the information for assembling the enzymes of a metabolic pathway.
a. A typical operon includes structural genes, a promoter region, an operator region, and a regulatory gene.
b. Structural genes code for the enzymes themselves. All structural genes are translated from one mRNA.
c. RNA polymerase binds at the promoter.
d. The operator is the site between the promoter and the first structural gene, where a regulatory protein can bind.
e. A repressor is an example of a gene regulatory protein.
f. Regulatory genes code for repressor proteins.
g. When the repressor is bound to an operator, transcription is switched off.
2. The lac operon:
a. The lac operon is an inducible operon; it is turned on in the presence of lactose.
(1) The lac operon contains three tandem structural genes.
(2) Lactose binds to the repressor, changing its conformation and making it unable to attach to the operator.
(3) Repressor protein can bind to the operator and prevent transcription from the lac operon only in the absence of lactose.
b. Positive control by cyclic AMP:
(1) The lac repressor exerts negative control.
(2) The glucose effect is an example of positive control.
(3) Cyclic AMP (cAMP) acts by binding to the catabolite gene activator protein (CAP).

(4) Binding of CAP-cAMP to the lac control region changes the conformation of DNA, which allows RNA polymerase to transcribe the lac operon.

3. The trp operon:
   a. The trp operon is a repressible operon; it is turned off in the presence of tryptophan.
   b. The trp operon repressor is active only when it is bound to a corepressor such as tryptophan.

IV. Control of Gene Expression in Eukaryotes

A. Introduction:
   1. Cells of a complex eukaryotic organism exist in many differentiated states.
      a. Differentiated cells retain a full set of genes, as demonstrated in experiments with plant root cells.
      b. Nuclei from cells of adult animals are capable of supporting the development of a new individual, as demonstrated in experiments with Xenopus oocytes.
   2. Genes are turned on or off as a result of interaction with regulatory proteins.
      a. Each cell type contains a unique set of proteins, characteristic of its state of differentiation.
      b. Regulation of gene expression occurs on three levels:
         (1) Transcriptional-level control mechanisms determine whether a gene will be transcribed, and how often.
         (2) Processing-level control mechanisms regulate the processing of primary RNA transcript into a messenger RNA.
         (3) Translational-level control mechanisms regulate the translation of mRNAs.

B. Transcriptional-level control:
   1. Differential transcription is the most important mechanism by which eukaryotic cells determine which proteins are synthesized.
   2. DNA microarrays monitor the expression of thousands of genes simultaneously.
      a. Immobilized fragments of DNA are hybridized with fluorescent cDNAs from cells under different conditions.
      b. Genes that are expressed show up as fluorescent spots on an array of immobilized genes.
      c. Microarrays provide a visual portrait of gene expression.
   3. The role of transcription factors in regulating gene expression:
      a. Transcription factors are the proteins that either stimulate or inhibit transcription by binding specific DNA sequences.
         (1) A single gene can be controlled by different regulatory proteins.
         (2) A single DNA-binding protein may control the expression of many different genes.
      b. Regulatory sequences are located upstream of the gene.
         (1) The TATA box is where general transcription factors assemble.
         (2) Basal level of gene transcription may also require two other promoter sequences, the CAAT box and GC box.
      c. PEPCK is a key enzyme in gluconeogenesis that is controlled by a variety of different transcription factors.
      d. Promoter regions are determined by deletion mapping and by DNA fingerprinting.
      e. The basal level of transcription is modified (either stimulated or inhibited) by other DNA-binding proteins.

(1) Binding of activated glucocorticoid receptor protein (in complex with hormone) to the glucocorticoid response elements (GRE) increases transcription.

(2) Glucocorticoids stimulate PEPCK gene expression.

4. The structure of transcription factors:
   a. Transcription factors contain a DNA binding domain and an activation domain.
      (1) The DNA-binding domains of most transcription factors have related structures (motifs) that interact with DNA sequences.
         (a) Most of the motifs contain a segment that binds to the major groove of the DNA.
         (b) The zinc finger, the helix-loop-helix, and the HMG box are the most common motifs.
      (2) In zinc finger proteins, the zinc ion of each finger is held in place by two cysteines and two histidines.
      (3) The helix-loop-helix (HLH) motif has two alpha-helical segments separated by a loop.
      (4) The leucine zipper motif has a leucine at every 7th amino acid of an alpha helix.
      (5) The high mobility group (HMG) box motif consists of three alpha helices.

5. DNA sites involved in regulating transcription:
   a. The TATA box regulates the initiation of transcription.
   b. The core promoter, from the TATA box to the start, is where the initiation complex assembles.
   c. The CAAT and the GC box are upstream and are required for initiation.
   d. All of the above constitute the proximal promoter elements.
   e. Researchers use deletion mapping, DNA footprinting, and genome-wide location analysis to find DNA sequences involved in regulation.

6. The Glucocorticoid Receptor (GR):
   a. The GR is a nuclear receptor that has three domains: DNA-binding domain, ligand-binding domain, and transcription-activator domain.
   b. The GR-binding sequence (GRE) is a palindrome.
   c. Pairs of GR molecules form dimers in which each subunit of the dimer binds to one-half of the DNA palindrome.
   d. Each DNA-binding domain of the GR contains two α–helical loops, each with a zinc ion at its base, called zinc fingers.

C. Transcriptional activation: The role of enhancers, promoters, and coactivators
   1. Enhancers are DNA elements that stimulate transcription above the basal level.
      a. Enhancers can be located very far upstream from the regulated gene.
      b. Transcription activators bind to the enhancer and stimulate transcription by either:
         (1) inducing changes in the chromatin that promote the assembly of the basal transcription machinery.
         (2) interacting with components of the basal transcription machinery.
   2. Coactivators, such as Mediator, interact with the basal transcription machinery.
   3. Coactivators that alter chromatin structure modify histones to regulate transcription.
      a. Histone acetyltransferases (HAT) maintains euchromatin and enhances transcription.

b. Chromatin remodeling complexes use ATP to enhance transcription.
D. Transcriptional Repression:
1. Histone deacetylases (HDACs) remove acetyl groups and repress transcription.
2. DNA methylation
a. DNA methylation in eukaryotic cells silences transcription.
b. DNA methylation mostly occurs in G-C-rich "islands," in or near transcriptional regulatory regions.
c. Methylation patterns of gene regulatory regions change during cellular differentiation.
3. Genomic imprinting:
a. Activity of certain genes, called imprinted genes, depends on whether they originated with the sperm or the egg.
b. Inactive and active versions of imprinted genes differ in their methylation patterns.

V. **Processing-level control:**
A. Protein diversity can be generated by alternative splicing.
B. There may be more than one way to process a primary transcript.
1. An intron can be spliced out of the transcript or retained.
2. An mRNA may contain alternate 5' or 3' terminal sequences.
C. The mechanism for selecting alternative splice sites depends on a splicing enhancer located within the exons whose inclusion is regulated.

VI. **Translational-level control:**
A. Translation of mRNAs that have been transported from nucleus to cytoplasm is regulated.
1. Translational-level control occurs via interactions of specific mRNAs and proteins in the cytoplasm.
2. Regulatory proteins act on untranslated regions (UTRs) at 5' and 3' ends of mRNA.
B. Cytoplasmic localization of mRNAs:
1. In the fruit fly embryo, the development of the anterior-posterior axis is regulated by the localization of specific mRNAs along this axis in the egg.
2. Cytoplasmic localization of mRNAs is determined by their 3' UTRs.
C. The control of mRNA translation:
1. In an unfertilized egg, mRNAs are stored in inactive form (masked mRNAs).
2. Masked mRNAs are associated with inhibitory proteins.
3. The mRNA translation rate can be controlled in different ways:
a. phosphorylation of initiation factors
b. binding of repressor molecules
D. The control of mRNA stability:
1. The lifetimes of eukaryotic mRNAs vary widely.
2. Poly(A) tail length may influence the longevity of mRNA.
a. The poly(A) tail is bound by a poly(A) binding protein (PABP).
b. PABP protects the tail from general nuclease activity.
c. PABP increases the sensitivity of the tail to a specific poly(A) ribonuclease.
d. While an mRNA is in cytoplasm, its tail its gradually shortened by the poly(A) nuclease.
e. When the tail is about 30 residues, it can no longer bind PABP; the mRNA is degraded.
3. Certain destabilizing sequences in the 3' UTR may affect the rate of poly(A) tail shortening.
4. Other mechanisms of translational-level control:

a. In translational frameshifting, a change of the reading frame by the ribosome generates two different polypeptides.

b. In a termination codon read through, the ribosome continues past a termination codon.

c. Editing of an mRNA after it has been transcribed occurs primarily in mitochondria.

d. Alternative translation initiation where different AUG codons in the same mRNA are used as the initiation codon.

e. In protein splicing, a segment can be excised from a specific polypeptide, and the two ends joined.

VI. **Posttranslational Control: Determining Protein Stability**

A. The protein profile of a cell depends on both control of synthesis and control of degradation.

1. Protein stability, measured as half-life, varies from minutes to weeks.

2. Protein stability may be determined by the amino acids on the N-terminus.

B. Proteins are degraded by barrel-shaped proteasomes.

1. Proteasomes occur in both cytosol and nuclei.

2. Proteasomes recognize proteins linked to ubiquitin.

VII. **Experimental Pathways on the Web: Genes That Control Embryonic Development**

A. Homeotic genes have been identified in fruit flies by the effect of mutations on the development of segmentation.

1. Mutations in homeogenes cause transformation of one body segment into another.

2. Different homeotic genes have a region of similarity called homeobox (Hox).

3. Hox sequences are present in a variety of eukaryotic organisms.

4. Hox sequences are in the coding region of genes and code for a homeodomain protein.

B. In mammals, the linear order of homeotic genes along the chromosome parallels the expression of those genes along the anterior-posterior axis of the embryo

C. Homeotic gene products function as transcription factors.

1. Homeotic proteins bind to specific DNA sequences.

2. The homeodomain itself determines the DNA binding specificity.

D. The eyeless gene in fruit flies controls development of the entire eye.

1. Genetically engineered flies in which eyeless could be controled developed ectopic eyes.

2. Mammals have a gene homologous to eyeless, indicating an evolutionary relationship in this homeobox gene.

Key Figure

Figure 12-28a. Organization of a bacterial operon.

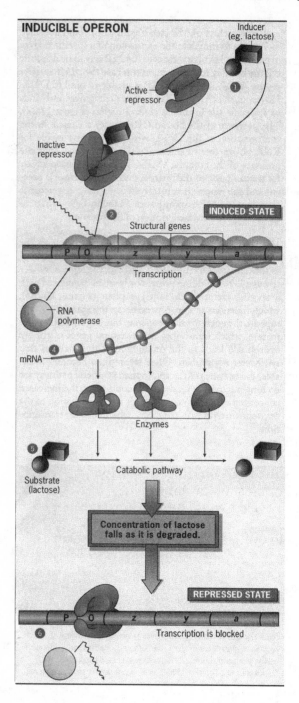

Questions for Thought:

1. In addition to the genes specifically shown in this diagram, what others must exist as part of the operon?

2. What features of this diagram identify it as a prokaryotic system?

3. This diagram shows induction of ß-galactosidase occurring at the level of transcription. Where else might gene expression be regulated?

4. If you knew only that ß-galactosidase was induced in the presence of lactose, but you didn't know whether the mechanism involved initiation of transcription or activation of existing enzyme molecules, what experiment could you perform to determine the exact mechanism of induction?

Review Problems

Short Answer

1. Give the significance (not the definition) of the following terms or phrases. Say what they
    do or why they are important. For example:

    lac operon: *enables prokaryotic cells to regulate the synthesis of enzymes for lactose metabolism
    only when it is appropriate to do so, i.e., when there is lactose in the environment.*

    a. nuclear pore complex

    b. nuclear localization signal

    c. evolutionary conservation of histone proteins

    d. H1 histone

    e. nuclear matrix

    f. selective gene amplification

    g. DNA rearrangement of antibody genes

    h. heterodimerization of transcription factors

    i. DNase I hypersensitive sites

    j. acetylated histones

2.   Compare and contrast the following:

   a.   heterochromatin vs. euchromatin

   b.   constitutive vs. facultative heterochromatin

   c.   repressible vs. inducible operons

   d.   DNA methylation in prokaryotes vs. DNA methylation in eukaryotes

   e.   liver fibronectin vs. fibroblast fibronectin

Multiple Choice

1.   Which of the following in not true of the nuclear envelope?
      a. The nuclear envelope is exactly like other cellular membranes.
      b. The nuclear envelope separates the genetic material from the cytoplasm.
      c. The nuclear envelope is a pair of concentric membranes.
      d. The nuclear envelope is studded with pores.

2.   Which cytoplasmic fibrils are most like the nuclear lamins?
      a. microtubules
      b. microfilaments
      c. intermediate filaments
      d. actomyosin

3.   What kind of molecules must pass between the nucleus and the cytoplasm?
      a. DNA
      b. protein
      c. lipids
      d. carbohydrates

4.   Which statement best characterizes the nuclear localization signal?
      a. The NLS is typically a small molecular weight metabolic intermediate.
      b. The NLS is a stretch of hydrophobic amino acids on a protein's N terminus.
      c. The NLS is one or two stretches of basic amino acids on a protein's C-terminus.
      d. The NLS is a steroid that binds to DNA.

5. Which sequence represents increasing levels of chromosomal organization, from most dispersed to most condensed?
    a. nucleosomes–30 nm filaments–supercoiled loops–mitotic chromosomes
    b. nucleosomes–supercoiled loops–30 nm filaments–mitotic chromosomes
    c. nucleosomes–30 nm filaments–mitotic chromosomes–supercoiled loops
    d. mitotic chromosomes–30 nm filaments–supercoiled loops–nucleosomes

6. When chromatin is treated with nonspecific nucleases, what is the length of the resulting pieces of DNA?
    a. random numbers of base pairs
    b. about 60 base pairs
    c. about 8 base pairs
    d. about 200 base pairs

7. Which of the following statements about histones is not true?
    a. Histones are very similar between species.
    b. Histones have many basic amino acids.
    c. Histones are rich in lysine and arginine.
    d. Each histone has one single gene that codes for it.

8. In mammalian cells, the DNA of the centromere is characteristic of:
    a. facultative heterochromatin.
    b. constitutive heterochromatin.
    c. euchromatin.
    d. dispersed chromatin.

9. Why are adult women genetic mosaics?
    a. Inactivated X chromosomes are all derived from the male parent.
    b. Inactivated X chromosomes are all derived from the female parent.
    c. Inactivated X chromosomes can be from either the male or the female parent.
    d. None of the above.

10. What do telomeres do?
    a. They protect the chromosomes from degradation by nucleases.
    b. They prevent the ends of chromosomes from fusing with one another.
    c. They are required for complete chromosomal replication.
    d. All of the above.

11. When all or a piece of a chromosome becomes attached to another chromosome, the aberration is called a(n):
    a. inversion.
    b. translocation.
    c. deletion.
    d. duplication.

12. If there were a mutation in the regulatory gene of an inducible promoter rendering the protein incapable of binding to the repressor, then:
    a. the structural genes would always be expressed.
    b. the structural genes would never be expressed.
    c. the structural genes would only be expressed in the presence of the inducer.
    d. the structural genes would only be expressed in the absence of the inducer.

13. Which of the following is a difference between inducible versus repressible operons?
    a. In an inducible operon, the ligand binds to the regulator protein.
    b. In a repressible operon, the ligand-regulator complex binds to the operator.
    c. In a repressible operon, all the structural gene products have related functions.
    d. In an inducible operon, RNA polymerase binds to the promoter.

14. To stimulate transcription, enhancer sequences:
    a. must be within a few base pairs of the gene they enhance.
    b. must be within a few hundred base pairs of the gene they enhance.
    c. can be tens of thousands of base pairs away from the genes they enhance.
    d. will not function if they are moved experimentally.

15. Which of the following is not a characteristic of transcription factors?
    a. Transcription factors have two fold rotational symmetry.
    b. Active transcription factors are dimers.
    c. Transcription factors are rich in basic amino acids.
    d. Transcription factors have one of a few DNA-binding motifs.

16. Eukaryotes regulate gene expression at all of the following levels except:
    a. transcription.
    b. translation.
    c. RNA processing.
    d. all of the above.

17. You are studying a mutant protein that has a half-life of several weeks. The normal, wild-type version has a half-life of only a few minutes. What feature would you look for to account for the difference in half-life of your mutant compared to the wild-type?
    a. misfolding during translation
    b. the presence of specific lysines, especially at the N-terminus
    c. the presence of a poly A tail
    d. none of the above.

18. Alternate splicing means that:
    a. the same gene can code for several different proteins.
    b. several different genes can code for the same protein.
    c. gene expression can be regulated at the level of transcription.
    d. pieces of DNA can move around within the genome.

19. Which of the following statements is true?
    a. A cell can potentially make fewer different proteins than the number of different genes it contains.
    b. A cell can potentially make only the same number of different proteins as the number of different genes it contains.
    c. A cell can potentially make more different proteins than the number of different genes it contains.
    d. None of the above.

20. The longevity of mRNA is related to:
    a. the length of the poly (A) tail.
    b. the presence of PABP.
    c. the nucleotide sequence of the 3′ UTR.
    d. all of the above.

Problems and Essays

1. The lac operon in *E. coli* has three regulatory regions (the gene for the repressor, i; the operator, O; and the promoter, P) and three structural genes (the gene for ß-galactosidase, z; the gene for galactoside permease, y; and the gene for a transacetylase, a).

In the absence of mutations, the structural genes, z, y, and a, are all expressed in the presence of lactose and repressed when lactose is not present. The genotype of the wild-type bacterium is

$$i^+ \ P^+ \ O^+ \ z^+ \ y^+ \ a^+$$

a. For the following mutations, would the structural genes be expressed in the presence of lactose? What about in the absence of lactose? The answer to the first one is given as an example.

Genotype	With Lactose			Without Lactose		
	ß-gal	gal permease	transacet	ß-gal	gal permease	transacet
$i^+ \ P^+ \ O^+ \ z^+ \ y^+ \ a^-$	yes	yes	no	no	no	no
$i^+ \ P^+ \ O^+ \ z^- \ y^+ \ a^+$						
$i^- \ P^+ \ O^+ \ z^+ \ y^+ \ a^+$						
$i^+ \ P^- \ O^+ \ z^+ \ y^+ \ a^+$						
$i^+ \ P^+ \ O^- \ z^+ \ y^+ \ a^+$						

b. What would happen to the expression of the structural genes in the presence and absence of lactose if there were a mutation in the promoter region of the i gene, P_i?

2. Chick DNA was isolated from both brain tissue and red blood cells and was treated with varying concentrations of DNase I. The DNA was then cut with a restriction enzyme that releases both the globin gene and the ovalbumin gene in fragments of characteristic size. These fragments were separated on agarose gels, and hybridized with radiolabeled probes for both genes. The results follow:

a. Neither tissue actively transcribed the ovalbumin gene. Which band, 1 or 2, most likely represents the fragment with the ovalbumin gene? Why?

b. Which gel, A or B, most likely represents the fragments from red blood cells? Why?

3. What specific problem is presented by the need for copious amounts of rRNA (e.g., in the oocytes of amphibians) as opposed to the need for protein? What strategies have cells evolved for providing rRNA?

4. Although many different transcription factors have been isolated, most can be grouped into one of a few families, according to recurring structural motifs. For the following proteins, indicate which structural motif is present.

_____This transcription factor occurs as a heterodimer with two alpha-helical segments separated by a loop. The α-helices are perpendicular to one another.

_____MyoD

_____This transcription factor has several zinc ions affiliated with it.

_____This transcription factor bends DNA, resulting in a spatial arrangement that enhances its access to other regulatory factors.

_____TFIIIA

_____UBF

5. Indicate whether the following describe transcriptionally active or inactive chromatin:

_____Core histones are heavily acetylated.

_____Chromatin is tightly affiliated with histones.

_____Chromatin is hypersensitive to DNase I.

_____Chromatin is methylated.

_____Chromatin is affiliated with histones, but the histone octamers are disrupted.

_____Chromatin is in the form of constitutive heterochromatin.

Web Explorations

When egg and sperm join at fertilization, each gamete contributes one copy—one allele—for each gene to the newly created zygote. For most genes, the effect that an allele has on the offspring is independent of its parental origin. An allele for blood type B, for example, would result in expression of the B type surface antigen on blood cells regardless of whether it came from the egg or the sperm. In the 1980s, however, some unusual genes were discovered that, unlike most, appear to "remember" where they came from. These were called imprinted genes. For imprinted genes, alleles of maternal origin act differently than those of paternal origin.

Genetic imprinting is an example of regulation of gene expression at the level of transcription. If, for example, only the maternal allele for a particular gene is expressed, then the paternal allele for that gene must be silenced. Likewise some alleles are only expressed when they are inherited paternally, and the maternal counterpart is transcriptionally silenced. Using the World Wide Web, find examples of imprinted genes in humans. For each example, learn which allele is expressed and which one is silenced. How are the alleles that are not expressed silenced? What disorders can result if an individual inherits both alleles of an imprinted gene from the same parent?

13   DNA Replication and Repair

Learning Objectives

When you have finished this chapter, you should be able to:

1. Define DNA replication. Describe the experiments in both prokaryotes and eukaryotes that established that DNA replication is semiconservative.
2. Give an overview of the replication process in prokaryotes, including the requirements for replication and the components of the replication complex.
3. Know the properties of different DNA polymerases, and the roles of DNA ligase, primase, DNA gyrase, helicase and single-stranded DNA-binding proteins.
4. Describe specific features of replication in eukaryotes. Compare and contrast these with specific features of replication in prokaryotes.
5. Recognize both the necessity for DNA repair and the various ways cells have evolved to repair DNA.
6. Describe some conditions caused by defects in the DNA repair mechanisms of humans. Explain how these conditions have been used to investigate the mechanisms of DNA repair.

Key Terms and Phrases

replication	semiconservative replication	conservative replication
dispersive replication	temperature-sensitive (ts) mutants	Meselson and Stahl
nonpermissive temperature	replication fork	origin
bidirectional replication	DNA polymerase	Arthur Kornberg
primer	semidiscontinuous replication	leading strand
lagging strand	Okazaki fragments	DNA ligase
primase	DNA gyrase	helicase
replisome	single-stranded DNA-binding protein (SSB)	
oriC site	initiator protein	ter site
replicase	exonuclease	replicon
recognition complex	autonomous replicating sequences (ARSs)	
nucleotide excision repair	base excision repair	mismatch repair
replication foci		

Reviewing the Chapter

I. Introduction
    A. Self-duplication is a fundamental property of life.
    B. DNA duplicates by a process called replication.
    C. DNA replication requires a number of auxiliary materials.
    D. The DNA replication machinery is also used for DNA repair.

II. DNA Replication
    A. DNA replication takes place by separation of the strands of the double helix, and synthesis of two daughter strands complementary to the two parental templates.

189

B. Semiconservative replication:
  1. During DNA replication, half of the parent structure is retained in each of the daughter duplexes.
  2. The Meselson and Stahl experiments supported the semiconservative model of replication in bacterial cells.
  3. Semiconservative replication was later demonstrated in eukaryotes.
C. Replication in bacterial cells:
  1. Temperature-sensitive mutants were used to identify the genes of replication.
  2. Replication can be studied using in vitro systems reconstituted from purified cellular compounds.
  3. Replication forks and bidirectional replication:
    a. Replication starts at the origin site.
    b. Replication proceeds bidirectionally.
    c. Replication forks are the points where a pair of replicating segments come together and join the nonreplicated segments.
  4. Unwinding the duplex and separating the strands:
    a. Tension is built up as DNA unwinds and the DNA ahead of the replication fork becomes positively supercoiled.
    b. DNA gyrase relieves the tension by changing the DNA into negatively supercoiled DNA.
  5. The properties of DNA polymerases:
    a DNA polymerase is DNA-dependent, i.e., it uses DNA as a template.
    B DNA polymerase requires a primer which provides the 3' hydroxyl terminus on which to add new nucleotides.
    c. Polymerization only occurs in a 5'-to-3' direction.
    d. There are three classes of DNA polymerases in bacterial cells. None is able to initiate DNA chains.
  6. Semidiscontinuous replication:
    a. Both daughter strands are synthesized simultaneously.
    b. The leading strand (in direction of the replication fork movement) is synthesized continuously.
    c. The lagging strand (in the opposite direction of the replication fork movement) is synthesized discontinuously.
    d. The lagging strand is constructed of small Okazaki fragments, which are joined by DNA ligase.
    e. Primase is an RNA polymerase that assembles short RNA primers; these primers are later removed and the gaps are sealed.
  7. The machinery operating at the replication fork:
    a. Helicase and single-stranded DNA-binding proteins unwind the parental duplex and separate the two strands.
    b. Primase and helicase form a primosome, which processively moves along the lagging-strand template.
    c. A single replisome synthesizes both leading and lagging strands.
  8. Initiation and termination of replication:
    a. In bacterial cells, replication starts at the oriC site.
    b. Replication begins with binding of an initiator protein.
    c. Replication terminates at the ter site.
D. The structure and functions of DNA polymerases:
  1. DNA polymerase III is the primary replication enzyme.
    a. DNA polymerase III holoenzyme contains ten different subunits having various functions in the replication process.
    b. By forming a sliding clamp, the ß-subunit maintains an association between the polymerase and the DNA template.

2. DNA polymerase I is involved in DNA repair and also removes RNA primers and replaces them with DNA.
3. The function of DNA polymerase II is unclear.
4. Exonuclease activities of DNA polymerases:
   a. Exonuclease degrades nucleic acids by removing the 5′ or 3′ terminal nucleotides one at a time.
   b. DNA polymerase I has 5′-to-3′ and 3′-to-5′ exonuclease activity; each exonuclease activity, as well as the polymerase activity is assigned to a separate domain of the enzyme.
   c. The 5′-to-3′ exonuclease of DNA polymerase I degrades RNA primers.
5. Ensuring high fidelity during DNA replication:
   a. The error rate of incorporation of an incorrect nucleotide during DNA replication is the spontaneous mutation rate.
   b. Acceptability of the incoming base is determined by the geometry of the base pair.
   c. During proofreading, mismatched bases are excised.
   d. Careful selection of the nucleotide, immediate proofreading, and postreplicative mismatch repair account for low error rates in replication (about 10^{-9}).
   e. Replication is rapid (about 1000 nucleotides per second).
E. Replication in eukaryotic cells:
   1. Replication in eukaryotes is less well understood than in prokaryotes.
      a. Mutant yeast cells that are unable to produce specific gene products are used to study replication.
      b. In vitro systems can be reconstituted with eukaryotic cellular extracts and/or purified components.
   2. Initiation of replication in eukaryotic cells:
      a. Eukaryote replicate their genome in small portions termed replicons from which replication proceeds outwards.
      b. Initiation of DNA synthesis in a replicon is regulated.
      c. Replicons located close together in a chromosome tend to undergo replication simultaneously.
      d. More tightly compacted regions are replicated later in the cell cycle.
      e. Origins of replication identified in yeast are called autonomous replicating sequences (ARSs).
      f. A multiprotein origin recognition complex (ORC) is assembled at the ARS. Mutation in the ARS prevents ORC from binding.
      g. Replication origins have not yet been identified in mammals.
   3. Restricting replication to once per cell cycle:
      a. Replication origins pass through different states so that they only replicate their DNA one time during a cell cycle.
         (1) Origin of replication is bound by an ORC.
         (2) Licensing factors bind to the origins.
         (3) Activation factors bind to the chromosomes and induce the licensed origins to initiate replication.
   4. The eukaryotic replication fork:
      a. Replication activities are similar in eukaryotes and prokaryotes.
      b. There are five different DNA polymerases in eukaryotes:
         (1) Polymerase α is associated with primase.
         (2) Polymerase δ is the major DNA-replicating enzyme.
         (3) Polymerase β is involved in DNA repair.
         (4) Polymerase γ replicates mitochondrial DNA.
      c. Eukaryotic DNA polymerases elongate in the 5′-to-3′ direction and require a primer. Most possess 3′-to-5′ exonuclease activity.

5. Replication and nuclear structure:
    a. The replication machinery is immobilized in the nuclear matrix.
    b. Replication forks are located within replication foci.
6. Chromatin structure and replication:
    a. DNA replication is associated with the opening of the nucleosome core particles.
    b. The assembly of DNA into nucleosomes is a very rapid event.
    c. Histones remain intact during replication and old and new histone octamers are distributed randomly between the two daughter duplexes.

III. DNA Repair
A. DNA repair is essential for cell survival.
    1. Cells are subjected to a variety of destructive forces.
    2. DNA is the cell molecule that is most susceptible to environmental damage.
    3. Ionizing radiation, common chemicals, ultraviolet radiation and thermal energy create various spontaneous alterations (lesions) in DNA.
    4. Cells have a number of mechanisms to repair genetic damage.
    5. Excision is the main mechanism of DNA repair
B. Nucleotide excision repair:
    1. A nucleotide excision repair (NER) system removes bulky lesions, such as pyrimidine dimers and chemically altered nucleotides.
    2. TFIIH is a key component of repair machinery and is also involved in the initiation of transcription. It links transcription and DNA repair.
    3. An endonuclease makes nicks on two sides of the altered strand, after which damaged oligonucleotides are removed by helicase and the gap is filled by DNA polymerase and sealed by DNA ligase.
    4. Nucleotide excision repair consists of two pathways:
        a. The preferential pathway selectively repairs the genes of greatest importance to the cell.
        b. A slower pathway repairs the rest of the genome.
C. Base excision repair:
    1. Base excision repair (BER) removes altered nucleotides that produce distortions of the double helix.
    2. DNA glycosylase recognizes the alteration and cleaves the base from the sugar.
    3. DNA glycosylases are specific for particular types of altered bases.
    4. Thymine occurs in DNA instead of uracil because otherwise the repair system would not have been able to distinguish between uracil and altered cytosine.
D. Mismatch repair:
    1. Mismatch repair corrects the mistakes that escape the DNA polymerase proofreading.
    2. Repair enzymes recognize distortions caused by mismatched bases.
    3. The repair system must distinguish between parental and new strands.
    4. In bacteria, the two strands are distinguished by the presence of methyl groups on the parental strand.
    5. The mechanism of mismatch repair in eukaryotes involves the presence of certain replication proteins that remain bound to the daughter strands for a time.

IV. The Human Perspective: Human Nucleotide Excision Repair Deficiencies
A. Excision repair deficiencies are the cause of a number of human disorders.
B. Xeroderma pigmentosum patients cannot repair sun damaged DNA.
C. Cockayne syndrome patients cannot repair transcriptionally active DNA.

D. Skin cancers due to sun-damaged DNA are increasing, including malignant melanoma, a potential killer.

E. Some colon cancers are due to mutations in mismatch repair genes.

V. Between Replication and Repair

A. DNA pol η can bypass thymine dimers during replication
1. DNA pol η acts in translesion synthesis (TLS).
2. DNA pol η is recruited to replicate short stretches of damaged DNA.
3. DNA pol η is in a superfamily of TLS polymerases.

B. Patients with mutant DNA pol η genes have a variant form of xeroderma pigmentosa.

VI. Experimental Pathways on the Web: The Role of Human NER Deficiencies in DNA Repair Research

A. DNA repair deficiency leads to an increased sensitivity to sunlight, in both bacteria and humans.

B. Cells of individuals afflicted with xeroderma pigmentosum are unable to properly excise pyrimidine dimers.

C. Several human genes involved in DNA repair have been isolated using intergenic complementation and transfection-correction techniques.

Key Figure

Figure 13.13   Replication of both the leading and lagging strands in *E. coli* is accomplished by DNA polymerases working together as part of a single complex.

Questions for Thought:

1. Why is it necessary for one of the template strands being replicated to fold back onto the DNA polymerase as the polymerase moves down the DNA?
2. What role do the SSB proteins play in this process? Does one strand require more SSBs than the other? Why or why not?
3. For simplicity, this figure does not show the torsional stress put on DNA by a replication complex, or the enzymes that relieve that stress. Which side of the replication fork shown here would experience the greatest stress?
4. What experimental evidence led to the discovery of Okazaki fragments?

Review Problems

Short Answer

1. Give the significance (not the definition) of the following terms or phrases. Say what they do or why they are important. For example:
   DNA gyrase: *relieves the tension caused by the unwinding of the DNA helix during replication.*

   a. Meselson and Stahl experiment

   b. temperature-sensitive mutants

   c. the exonuclease activity of DNA polymerase

   d. RNA primers

   e. single-stranded binding proteins

   f. oriC site

   g. autonomous replicating sequences (ARSs)

   h. DNA repair enzymes

   i. intergenic complementation technique

2. Compare and contrast the following:

   a. RNA primers of leading strands vs. RNA primers of lagging strands

   b. oriC vs. ter

   c. base substitution errors in transcription vs. base substitutions in replication

   d. replication origins in prokaryotes vs. replication origins in eukaryotes

   e. autonomous replicating sequence vs. origin recognition complex

Multiple Choice

1. Semiconservative replication of DNA means that:
   a. each daughter duplex will have one of the original parental strands and one new strand.
   b. one daughter duplex will be entirely new and the other will have both original parental strands.
   c. both daughter duplexes will be entirely new and the parental duplex will be degraded.
   d. each strand of each daughter duplex will have parts of the parental strands and parts of new strands.

2. In vitro systems used to study replication must include all of the following except:
   a. a DNA molecule to be replicated.
   b. deoxynucleotide triphosphates.
   c. isolated cellular nuclei.
   d. cellular extracts with the factors and cofactors of replication.

3. Strain introduced by unwinding DNA at the replication forks is relieved by
   a. DNA gyrase.
   b. type I topoisomerases.
   c. type II topoisomerases.
   d. all of the above.

4. A replicating prokaryotic chromosome has __?__ replication forks, and a replicating eukaryotic chromosome has __?__.
   a. one; many
   b. two; two
   c. two; many
   d. many; many

5. Which of the following forms of DNA can serve as a template for DNA polymerase?
   a. partially double-standed DNA
   b. circular double-stranded DNA
   c. intact double-standed DNA
   d. circular single-stranded DNA

6. If replicating bacteria are given a short pulse of radiolabeled thymidine and rapidly killed, the radiolabel shows up as:
   a. large DNA molecules only.
   b. mostly large DNA molecules and very few short DNA molecules.
   c. mostly short DNA molecules and very few large DNA molecules.
   d. free nucleotides only.

7. An input of energy is not requried for which of the following steps of replication?
   a. separating the two DNA strands.
   b. unwinding the coiled DNA.
   c. linking together the individual deoxynucleotides.
   d. All of the above require energy.

8. If a reaction mixture of replicating DNA is suddenly diluted:
   a. the rate of DNA synthesis is slowed.
   b. the rate of DNA synthesis is unchanged.
   c. the rate of DNA synthesis is speeded up.
   d. DNA synthesis stops.

9. Experiments such as that described in number (8) above demonstrated that:
   a. a new DNA polymerase molecule enters the complex to synthesize each Okazaki fragment.
   b. the same DNA polymerase molecule synthesizes each new Okazaki fragment.
   c. DNA polymerase molecules are not involved in DNA synthesis.
   d. none of the above

10. The replisome, which synthesizes DNA on both leading and lagging strands, always moves toward the replication fork. This must mean that:
    a. synthesis on the lagging strand requires that the strand fold back onto the replisome.
    b. synthesis on the lagging strand occurs in the 3'-to-5' direction, and synthesis on the leading strand occurs in the 5'-to-3' direction.
    c. synthesis on the leading strand occurs in the 3' to 5' direction, and synthesis on the lagging strand occurs in the 5'-to 3'-direction.
    d. DNA strands are lined up in the same direction, with 3' ends together and 5' ends together.

11. The exonuclease activity of the DNA polymerases functions to:
    a. remove the RNA primer sequences.
    b. proofread the new DNA strand and remove inappropriate nucleotides.
    c. maximize the fidelity of DNA replication.
    d. all of the above

12. Incorrect pairing of bases during bacterial DNA replication occurs about once every 10^5 to 10^6 nucleotides, but the spontaneous mutation rate is only one error in 10^9 base pairs. What accounts for the difference?
    a. proofreading and correcting by the DNA polymerase
    b. mismatch repair following replication
    c. both (a) and (b)
    d. neither (a) nor (b)

13. The last DNA to be replicated in the eukaryotic chromosome is:
    a. telomeres at the end of the chromosomes.
    b. heterochromatin.
    c. euchromatin in the arms of the chromosomes.
    d. facultative heterochromatin.

14. OriC is to *E. coli* as___?___is to yeast.
    a. autonomous replicating sequence
    b. origin recognition complex
    c. heterochromatin
    d. DNA polymerase δ

15. DNA polymerase III is to prokaryotes as ___?___ is to eukaryotes.
    a. DNA polymerase α
    b. DNA polymerase β
    c. DNA polymerase γ
    d. DNA polymerase δ

16. A newly replicated DNA duplex is likely to have:
    a. nucleosomes both from the parental DNA and newly assembled.
    b. nucleosomes, each with histones both from the parental nucleosomes and newly synthesized.
    c. nucleosomes either all newly synthesized or all from the parental strand.
    d. a random conbination of both nucleosomes and individual histones, both newly synthesized and from the parental strands.

17. How are damaged nucleotides recognized by DNA repair enzymes?
    a. They are tagged with methyl groups.
    b. They are tagged with acetyl groups.
    c. They are recognzied by their bulky geometry.
    d. They are recognized by lesion receptors.

18. If the genes from the following list were damaged, which gene would most likely be repaired first?
    a. the globin genes in a cell destined to become a chick red blood cell
    b. the heterochromatin in a liver cell
    c. the ovalbumin genes in a cell destined to become a chick red blood cell
    d. the antibody genes in a liver cell

19. Fibroblast cells from several different xeroderma pigmentosa patients were cultured, then fused in intergenic complemetation studies. Which cell fusions would not result in a hybrid capable of repairing damaged DNA?
    a. XP-A and XP-G hybrids
    b. XP-B and XP-B hybrids
    c. XP-C and XP-D hybrids
    d. XP-E and XP-F hybrids

20. Why should patients with xeroderma pigmentosum avoid sunlight?
   a. The UV wavelengths do irreparable damage to DNA.
   b. Sunlight inhibits any residual DNA repair activity in the cell.
   c. These patients lack pigmentation to protect them from burning.
   d. Sunlight inhibits DNA polymerases.

Problems and Essays

1. Onion root tip is a good source of actively growing cells undergoing DNA replication. Suppose you pulse-label some root tip cells with ^{3}H-thymidine, then isolate the chromosomes during metaphase of cell division. Using autoradiography, you examine the pattern of thymidine incorporation into the chromosomes.

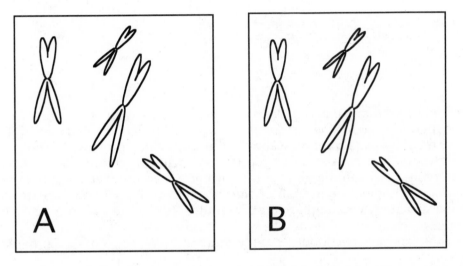

   a. In diagram (a), use dots to indicate where the radiolabeled thymidine appears during the first cell division.
   b. Show the pattern of radioactive labeling during the second cell division on diagram (b).
   c. Explain the labeling pattern that you put in each diagram.

2. On the following diagram of an E. coli replicon, draw leading strands, lagging strands, Okazaki fragments, and primers. Label (or color code) the parts that are DNA and the parts that are RNA. Also label the OriC site.

3. You have isolated a temperature-sensitive mutant of *E. coli* that can replicate its DNA at 20°C, but not at 37 °C. You are interested in identifying some of the important enzymes involved in DNA replication in normal cells. Using both your temperature-sensitive mutant and your cultures of normal *E. coli*, design an experiment to identify one such enzyme.

4. The fictional bacterium *Abra cadabra* has a chromosome with one origin of replication, OriC, and one digeston site for the restriction endonuclease, EcoR1. You have digested many copies of the *Abra cadabra* chromosome with EcoR1, then provided the linear chromosomes with all the raw materials for DNA replication, including cellular extracts. After various time intervals, you examine some of the chromosomes using electron microscopy. The structures you find before restriction digestion and at time= 0 minutes are shown below. Draw the structures you might find at 5 minutes, 10 minutes, 20 minutes, and 40 minutes.

a. Before restriction enzyme digestion: d. time= 10 mins

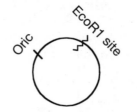

b. time= 0 mins   e. time= 20 mins

c. time= 5 mins   f. time= 40 mins

5. One frequent type of DNA lesion involves deamination of the deoxyribonucleotide bases. Next to each DNA base illustrated below, draw the structure (if any) that results from spontaneous deamination.

Natural Bases   **Deaminated Bases**   **Natural Bases**   **Deaminated Bases**

cytosine

guanine

thymine

adenine

Do any of your structures look familiar?

6. At least 20 different DNA glycosylases recognize and remove specific deaminated DNA bases. How would this important system of DNA repair be affected if uracil, rather than thymine, occurred in DNA as it does in RNA?

1

Web Explorations

In addition to xeroderma pigmentosum and XP-V, several other congenital disorders are related to failure of DNA repair systems. Surprisingly not all are linked to cancer, and not all patients suffering from these disorders are light sensitive. Some examples include Cockayne syndrome and trichothiodystrophy (TTD).

Using the World Wide Web, find out more about these DNA repair system disorders. What are the symptoms of these disorders and when do they appear? How are these congenital disorders explained on the molecular level, and what parts of the DNA repair machinery are affected?

14 Cellular Reproduction

Learning Objectives

When you have finished this chapter, you should be able to:

1. Define cellular reproduction and describe its main characteristics.
2. Define the cell cycle, including the different phases. Know in which phases the synthesis of major macromolecules occurs.
3. Understand how the cell cycle is controlled.
   a. Describe the role of cyclins and protein kinases in the cell cycle regulation.
   b. Describe the model systems used for studying the cell cycle, and recognize what makes them particularly suitable for such studies.
4. Give an overview of mitosis and meiosis, and summarize the differences between the two.
   a. Describe the formation of the mitotic spindle and explain how the chromosomes are moved during anaphase.
   b. Understand the difference between chromosomes and sister chromatids.
   c. Define centromere and kinetochore.
5. Describe prophase I of meiosis; define synaptonemal complex, chiasma, bivalent, and tetrad.
6. Understand genetic recombination and its importance in evolution.

Key Terms and Phrases

cell division
cellular reproduction
interphase
G_2
cyclin-dependent kinase (Cdk)
mitosis
centromere
mitotic spindle
prometaphase
anaphase A
cytokinesis
meiosis
primary spermatocytes
primary oocytes
sporophyte
meiosis I
prophase I
synapsis
pachytene
chiasma
recombination nodules

mother cell
cell cycle
G_1
maturation-promoting factor (MPF)
checkpoint
prophase
kinetochore
nuclear envelope breakdown
metaphase
anaphase B
contractile ring theory
gametic meiosis
spermatids
zygotic meiosis
sporogenesis
meiosis II
leptotene
synaptonemal complex (SC)
diplotene
mitotic nondisjunction
genetic recombination

daughter cell
M phase
S phase
cyclin
p53
chromatid
centrosome cycle
aster
anaphase
telophase
phragmoplast
spermatogonia
oogonia
sporic meiosis
gametophyte
interkinesis
zygotene
bivalent (tetrad)
diakinesis
aneuploidy
Holliday structure

Reviewing the Chapter

I. Introduction

    A. Cells reproduce by the process of cell division.

    B. A dividing cell is called a mother cell; its descendants are called daughter cells.

    C. The mother cell transmits copies of its genetic information to its daughter cells.

    D. Cellular gametes are formed by cell division.

II. The Cell Cycle

    A. The cell cycle is the succession of events from one cell division to another.

        1. The cell cycle consists of the M phase and interphase.

            a. The M phase includes the processes of mitosis and cytokinesis.

            b. The M phase usually lasts 30 to 60 minutes.

            c. Interphase, constituting the majority of the cell cycle, lasts longer than the M phase.

            d. Both synthesis of macromolecules and DNA replication occur during interphase.

        2. Interphase includes G_1, S, and G_2 periods.

            a. G_1 is the period between the end of mitosis and the beginning of DNA replication.

            b. DNA replication occurs during the S phase.

            c. The G_2 occurs between the end of DNA synthesis and the beginning of mitosis.

    B. Cell cycles in vivo:

        1. Three cell types are distinguished based on their capacity to grow and divide.

            a. Extremely specialized cells (nerve cells) have lost the ability to divide.

            b. Cells that normally do not divide can be induced to begin DNA synthesis (liver cells).

            c. Cells that have a high level of mitotic activity (stem cells).

        2. There is great variability in the length of the cell cycle in different cells.

            a. G_1 is most variable in length.

            b. Cells that stay indefinitely in G_1 are considered to be in the G_0 state.

    C. Control of the cell cycle:

        1. Cell cycle control is focused on initiation of DNA replication and initiation of mitosis.

        2. The cytoplasm contains factors that regulate the state of the nucleus, as evidenced from cell fusion experiments.

            a. The cytoplasm of a replicating cell contains factors that stimulate initiation of DNA synthesis.

            b. The cytoplasm of a cell undergoing mitosis contains factors that trigger chromosomal condensation.

            c. G_1-S and G_2-M transitions are both under positive control.

        3. The role of protein kinases:

            a. Entry of the cell into M phase is triggered by activation of a protein kinase called maturation-promoting factor (MPF).

                (1) MPF consists of two subunits: a kinase and a regulatory subunit, cyclin.

                (2) Increased concentration of cyclin activates the kinase.

                (3) The cyclin level fluctuates predictably during the cell cycle.

            b. MPF-like, cyclin-dependent kinases (Cdks) occur in yeast cells.

                (1) The product of the cdc2 gene in fission yeast and CDC28 in budding yeast is a cyclin-dependent kinase responsible for passage through both control points.

(2) For cells to pass through a point of commitment Cdks must be transiently activated by specific cyclins.
(3) Activated Cdk phosphorylates the sets of proteins specific for the particular transition.
 c. G_1 cyclins are responsible for G_1-S transition (START point).
 d. Mitotic cyclins are responsible for the G_2-M transition.
 e. Cyclins and cyclin-dependent kinases are highly conserved between species.
4. Cyclin concentration:
 a. Cyclin concentrations change by regulated transcription.
 b. Different cyclins are transcribed at different points in the cell cycle.
 c. Cyclin-Cdk complexes phosphorylate other proteins.
5. Cyclin binding:
 a. Cylcin binds to the catalytic subunit of Cdk.
 b. Cyclin binding reveals the enzyme's active site.
6. Cdk phosphorylation state:
 a. Cdc-activating kinase (CAK) phosphorylates both a threonine and a tyrosine on the Cdk subunit.
 b. The doubly phosphorylated Cdc-cyclin is inactive.
 c. A phosphatase (cdc25) removes one phosphate, at the tyrosine.
 d. The singly phosphorylated Cdc-cyclin is active, driving the cell to mitosis.
 e. CAK and cdc25 are activated by other kinases and phosphatases.
 f. Multilayered enzymes provide several targets for cell cycle regulation.
7. Cdk inhibitors, such as Sic1, also control cell cycle progression.
8. Controlled proteolysis of regulatory proteins occurs via the ubiquitin-proteasome pathway.
9. Subcellular localization: Movement of cylcins between the cytoplasm and the nucleus is another point of control.
10. Checkpoints, kinase inhibitors and cellular responses:
 a. Cells are able to control their progress through the cell cycle according to internal and external cues.
 b. Progress through the cell cycle can be arrested at a checkpoint by:
  (1) sensors that detect chromosomal abnormalities.
  (2) transmitters that signal the information.
  (3) effectors that inhibit cell cycle machinery.
 c. Pathways that stop cell cycle progress:
  (1) Chk1 inhibits Cdc25 when DNA is damaged.
  (2) p21 inhibits the G_1 Cdk when DNA is damaged.
 d. Cells may synthesize other proteins that inhibit progression through the cell cycle.

III. M Phase: Mitosis and Cytokinesis

A. Mitosis is a process of nuclear division in which two nuclei with identical genetic content are produced.
1. Mitosis is usually accompanied by cytokinesis.
2. Cytokinesis partitions the cytoplasm into two roughly equal parts.
3. Mitosis maintains the chromosome number.
4. Mitosis can occur in either diploid or haploid cells.
5. Mitosis consists of prophase, prometaphase, metaphase, anaphase, and telophase.
B. Prophase:
1. In prophase, duplicated chromosomes are prepared for segregation and mitotic machinery is assembled.
2. Formation of the mitotic chromosome:

a. Chromosome compaction or condensation occurs in early prophase.
    (1) Phosphorylation of histone H3 is required for compaction.
    (2) Condensin protein is also responsible for compaction.
        (a) Condensin is phophorylated by Cdk.
        (b) Phosphorylated condensin actively compacts DNA.
    (3) Type II topoisomerase is also necessary to supercoil DNA.
b. Compact chromosomes appears as rod-like structures.
    (1) Each mitotic chromosome consists of two chromatids.
    (2) Chromatids are connected to each other at their centromeres.
    (3) Proteins that mediate packaging of mitotic chromosomes form a scaffold to which DNA loops are attached.
    (4) Chromosome compaction requires the DNA-untangling enzyme topoisomerase II.
c. Centromeres and kinetochores:
    (1) Centromeres occur at primary constrictions on mitotic chromosomes and are the binding sites for proteins.
    (2) Kinetochores are on the outer surfaces of centromeres.
    (3) Kinetochores are the sites where chromosomes attach to the microtubules of the mitotic spindle.
3. Formation of the mitotic spindle:
  a. The mitotic spindle is made of microtubules.
  b. The centrosome cycle (doubling of centrioles in the cytoplasm) progresses simultaneously with the cell cycle.
  c. In the mitotic spindles of animal cells, microtubules are arranged in an aster formation around each centrosome.
    (1) Microtubules of the aster emerge from the pericentriolar material, which acts as a nucleating center.
    (2) Certain types of animal cells and most of higher plants lack centrosomes.
4. The dissolution of the nuclear envelope and the fragmentation of cytoplasmic organelles:
  a. Nuclear envelope breakdown is caused by Cdk kinase.
  b. Cdk kinase phosphorylates nuclear lamins, which leads to the disassembly of the nuclear lamina.
  c. The endoplasmic reticulum and Golgi complex are fragmented during mitosis, while mitochondria, lysosomes, and chloroplasts stay intact.

C. Prometaphase:
  1. During prometaphase the definitive mitotic spindle is formed and chromosomes are moved by microtubules into the center of the cell.
    a. The microtubules grow into the region around a chromosome
    b. The ends of the microtubules penetrate into the body of the kinetochore.
    c. A single kinetochore becomes attached to microtubules from both spindle poles.
  2. Chromosomes arrive at the center of the cell by differential growth of microtubules at opposite poles.

D. Metaphase:
  1. In metaphase, chromosomes are aligned at the spindle equator on the metaphase plane.
  2. Microtubules in a mitotic spindle are highly organized.
    a. Astral microtubules radiate from the centrosome to the region outside the body of the spindle and determine the plane of cytokinesis.

      b. Chromosomal (or kinetochore) microtubules extend from the centrosome to the kinetochores of the chromosomes; they move chromosomes to the poles.

      c. Polar (interpolar) microtubules extend from the centrosome past the chromosomes and maintain the integrity of the spindle.

   3. Microtubule flux in the metaphase spindle involves tubulin "treadmilling" toward the poles.

E. Anaphase begins when sister chromatids split and begin to move poleward.

   1. Control of anaphase:

      a. Anaphase promoting complex (APC) is activated at the metaphase/anaphase transition.

      b. An APC subunit marks an anaphase inhibitor, called securin, for destruction.

   2. The events of anaphase:

      a. Chromosomes are split in synchrony; chromatids begin migrating to the poles.

      b. As chromosomes move toward a pole, microtubules attached to kinetochores are shortened.

      c. Movement of chromosomes toward the poles is called anaphase A.

      d. Two spindle poles move in opposite directions due to elongation of polar microtubules; this process is called anaphase B.

   3. Forces required for chromosome movement at anaphase:

      a. Both dynein and kinesin are found at kinetochores of chromosomes.

      b. Microtubule disassembly generates sufficient force to move chromosomes.

F. The spindle checkpoint: The misalignment of one or several chromosomes delays chromosome separation.

   1. MAD2 gene is mutant in misaligned cells.

   2. Cell cycle arrest occurs when a cell contains a chromosome that is attached by microtubules to only one spindle pole.

G. Telophase:

   1. The cell returns to interphase condition during telophase.

      a. Nuclear envelopes of two nuclei are reassembled.

      b. Chromosomes become dispersed.

      c. The endoplasmic reticulum and Golgi complex re-form.

   2. The cytoplasm is partitioned into two cells.

H. Forces required for mitotic movements:

   1. Mitotic movement is powered by microtubule motors (dynein and kinesin-related proteins).

      a. Microtubule motors are located at the spindle poles and kinetochores.

      b. Motor proteins can be studied in three ways:

         (1) by analyzing a mutant cell without a motor.

         (2) by using antibodies against motor proteins.

         (3) by depletion of motor proteins from cell extracts with spindles.

      c. Motor proteins have a number of general features.

         (1) Motor proteins along polar microtubules keep the poles apart.

         (2) Motor proteins residing on the chromosome bring chromosomes to the metaphase plate and keep them there.

         (3) Motor proteins in the region of spindle equator elongate the spindle during anaphase B.

I. Cytokinesis:

   1. Cytokinesis in animal cells:

a. Cytokinesis starts with an indentation of the cell surface, lying in the same plane as chromosomes of the metaphase plate.
b. The contractile ring theory suggested that a thin cortical band composed of actin and myosin filaments generates the force to cleave the cell.
   (1) Sliding of actin filaments pulls the cortex and attached plasma membrane to the center of the cell.
   (2) The cortical ring is assembled rapidly prior to cytokinesis and dismantled immediately after it.
   (3) The cortical ring forms midway between two poles and a signal emanates from the central part of the spindle.
c. Dissociation of both mitosis and cytokinesis is possible.
2. Cytokinesis in plant cells: Formation of the cell plate:
a. Plant cells build a cell membrane and cell wall in the cell center.
b. The cell plate begins with the appearance of the microtubular phragmoplast.
c. Material for the cell wall is brought to the phragmoplast by Golgi vesicles.
d. Cell plate formation proceeds laterally from the center of the cell.

IV. Meiosis

A. During meiosis, chromosome number is halved and haploid cells are formed
1. Each daughter cell has one member of each pair of homologous chromosomes.
2. Meiosis consists of two subsequent divisions.
a. In the first division, homologous chromosomes pair and then segregate ensuring that daughter cells receive a full haploid complement of chromosomes.
b. In the second division, the two chromatids are separated.
c. Genetic recombination occurs during pairing of homologues.
3. In different eukaryotes meiosis occurs at different points in the life cycle.
a. In gametic (terminal) meiosis, meiosis is linked to gamete formation.
   (1) Gametic meiosis is typical for all multicellular animals.
   (2) In vertebrates, formation of spermatozoa from spermatogonia and eggs from oogonia takes place by gametic meiosis.
b. In zygotic (initial) meiosis, meiosis occurs just after fertilization; zygotic meiosis occurs only in protists and fungi.
c. In sporic (intermediate) meiosis, meiosis is independent of gamete formation and fertilization.
   (1) Sporic meiosis is typical for higher plants.
   (2) Sporogenesis (production of spores) includes meiosis.
   (3) Gametes are produced by haploid gametophytes via mitosis.
B. The stages of meiosis:
1. DNA is replicated prior to meiosis. The premeiotic S phase lasts longer than a premitotic S phase.
2. Prophase I consists of several stages:
a. In leptotene, the first stage of prophase, chromosomal condensation starts.
b. During zygotene, the second stage of prophase I, homologous chromosomes pair.
   (1) Genetic recombination begins before the chromosomes are visibly paired.
   (2) In synapsis, homologues associate via the synaptonemal complex.
      (a) The synaptonemal complex is a ladderlike structure.

(b) The synaptonemal complex functions as a scaffold to allow interacting chromatids to complete their cross-over activities.

(c) A pair of synapsed homologous chromosomes form a bivalent (tetrad) composed of four chromatids.

   c. In pachytene, synapsis ends.
     (1) Recombination nodules appear on chromosomes.
     (2) Pachytene may last for weeks.

   d. During diplotene, the synaptonemal complex is disassembled and homologous chromosomes start moving apart.
     (1) Chiasmata are the remaining points of attachment between homologous chromosomes.
     (2) Chiasmata visually portray the extent of recombination.
     (3) Diplotene chromosomes can be observed as lampbrush chromosomes with characteristic loops of transcriptionally active DNA.

   e. During diakinesis, the last stage of prophase I, chromosomes are prepared for attachment to the meiotic spindle fibers.
     (1) Diakinesis ends with the disappearance of the nucleolus and the disassembly of the nuclear envelope.
     (2) Diakinesis is triggered by an increase in MPF activity.

3. During metaphase I, homologous chromosomes are aligned at the metaphase plate so that both chromatids of one chromosome face the same pole.
   a. Homologous chromosomes are held by one or several chiasmata.
   b. Absence of a chiasma can lead to abnormal segregation of chromosomes.

4. Homologous chromosomes separate during anaphase I.
   a. Maternal and paternal chromosomes of each tetrad segregate into the two daughter cells independent of other chromosomes.
   b. Behavior of chromosomes during anaphase I corresponds to Mendel's law of independent assortment.

5. Interkinesis is a short stage between two meiotic divisions.
   a. Cells at interkinesis have a haploid number of chromosomes.
   b. Cells have a diploid amount of nuclear DNA.

6. Meiosis II is simpler than meiosis I.
   a. During metaphase II, chromosomes are aligned so that the kinetochores of sister chromatids face opposite poles.
   b. Sister chromatids separate during anaphase II.
   c. Meiosis II produces cells haploid in both the amount of nuclear DNA and chromosome number.

C. The Human Perspective: Meiotic nondisjunction and its consequences:
   1. Meiotic nondisjunction occurs when homologous chromosomes do not separate during meiosis I or sister chromatids don't separate during meiosis II.
   2. Meiotic nondisjunction leads to the formation of gametes (and later zygotes) with abnormal numbers of chromosomes.
   3. Aneuploidy is a condition in which an abnormal number of chromosomes occurs in cells.
     a. Trisomy is the presence of one extra chromosome.
     b. Monosomy is the lack of one chromosome.
   4. Persons with an abnormal number of somatic chromosomes seldom survive.
   5. Persons with an abnormal number of sex chromosomes are sterile.
   6. A reduction in genetic recombination increases the risk of meiotic nondisjunction.

a. Genetic recombination is reduced in the oocytes of older women.

b. Infants of older mothers have a higher incidence of aneuploidy.

D. Genetic recombination during meiosis:

1. Meiosis increases genetic variability by mixing maternal and paternal alleles between homologous chromosomes.

2. According to early views, recombination took place by the actual physical breakage and reunion of an entire compacted chromatid.

3. According to modern views, recombination occurs via breakage and reunion of single DNA molecules.

   a. Recombination occurs without the addition or loss of a single base pair.

   b. Because of compaction, the breakage and reunion of an entire chromosome would randomly and irretrievably split hundreds of genes.

   c. Prior to recombination, DNA strands are aligned via homology search.

      (1) Breaks are introduced into one strand of each duplex at corresponding sites.

      (2) Broken strands invade the other DNA molecule.

      (3) The two duplexes are joined to each other by a Holliday structure.

   d. The point of crossover moves via branch migration.

4. Recombination intermediates in bacterial and viral genomes can be observed using the electron microscope.

5. Recombination in eukaryotes is more complex than in prokaryotes.

V. Experimental Pathways: The Discovery and Characterization of MPF

A. MPF was first observed in studies of the effect of cytoplasm on the state of the nucleus in oocytes.

B. In developing embryos, MPF activity fluctuates following the stages of the cell cycle.

C. Levels of cyclin were found to correlate with levels of MPF activity.

D. Purified MPF was shown to possess kinase activity and stimulate nuclei to prepare for entry into mitosis.

Key Figure

Figure 14.1. The eukaryotic cell cycle.

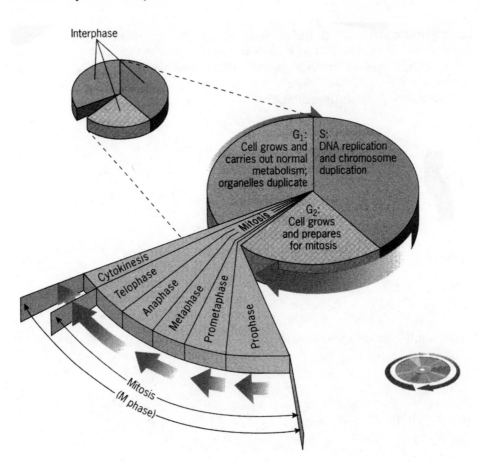

Questions for Thought:

1. In which phase of the cycle is protein synthesis occurring?
2. How would this diagram differ if it portrayed the cell cycle of a primary oocyte or primary spermatocyte? Why?
3. How would this diagram differ if it portrayed the cell cycle of a fully mature nerve cell? Why?
4. If the cycle is compared to a clock with the G_1-S transition at 12:00, at what hours are cylcins at their highest level?

Review Problems

1. Give the significance (not the definition) of the following terms or phrases. Say what they do or why they are important. For example:

    meiosis: *reduces the chromosome number by half, a necessary step for species that undergo sexual reproduction via fertilization.*

    a. MPF (maturation promoting factor)

    b. cell cycle checkpoints

    c. p53 protein

    d. formation of the mitotic chromosome

    e. phosphorylation of histones

    f. mitotic spindle

    g. fragmentation of the endoplasmic reticulum

    h. polar microtubules

    i. slow rate of chromosome movement during anaphase

    j. chiasmata

2. Compare and contrast the following:
   a. G_o vs. G_1 periods of the cell cycle

   b. centromere vs. kinetochore

   c. prometaphase vs. metaphase

   d. anaphase A vs. anaphase B

   e. phragmoplast vs. contractile ring

Multiple Choice

1. A cell that is entering the M phase of the cell cycle is:
   a. always diploid, either with duplicated or unduplicated chromosomes.
   b. always haploid, always with duplicated chromosomes.
   c. either haploid or diploid, always with duplicated chromosomes.
   d. either haploid or diploid, either with duplicated or unduplicated chromosomes.

2. A culture of cells is given ^{3}H-thymidine for 30 minutes; then a sample of the cells is fixed and examined by autoradiography. It is found that:
   a. all of the cells engaged in mitosis have incorporated the labeled thymidine into their chromosomes.
   b. none of the cells engaged in mitosis have incorporated the labeled thymidine into their chromosomes.
   c. all of the cells, both mitotic and nonmitotic, have incorporated the labeled thymidine into their chromosomes.
   d. none of the cells, neither mitotic nor nonmitotic, have incorporated the labeled thymidine into their chromosomes.

3. During which phases of the cell cycle do the chromosomes exist in a doubled state?
   a. G_2 and early M
   b. G_1 and S
   c. late M and G_1
   d. G_o and G_1

4. Liver cells are highly specialized, but when the liver is damaged or parts of it are removed surgically, the tissue grows. Liver cells fall into which category of cells?
   a. cells permanently locked in G_0
   b. cells that can be induced to enter the S phase
   c. cells subject to continual renewal
   d. cannot be determined from the information given

5. Throughout interphase,__?__are synthesized at a relatively constant rate, whereas__?__are synthesized only during the S phase.
   a. most major proteins, DNA replication enzymes
   b. histones, DNA replication enzymes
   c. structural proteins, functional proteins
   d. most major proteins, histones

6. If a cell in the G_1 phase is fused with a cell in the S phase, then
   a. the G_1 nucleus will enter the S phase.
   b. the S nucleus will enter the G_1 phase.
   c. both nuclei will enter the G_2 phase.
   d. both nuclei will shut down.

7. Cyclin concentrations are highest during which periods of the cell cycle?
   a. late G_1 and early S
   b. late G_2 and early M
   c. late G_1 and late G_2
   d. late M and late S

8. Cyclin is to MPF as__?__ is to cyclin-dependent kinases.
   a. cyclin
   b. MPF
   c. cdc2
   d. activator

9. The product of the cdc2 gene is
   a. cyclin.
   b. p53.
   c. cyclin-dependent kinase.
   d. CAK.

10. A mutation in the gene that codes for p53 can result in
    a. loss of an inhibitor protein, p21.
    b. loss of a transcription factor.
    c. uncontrolled cell growth and cancer.
    d. all of the above.

11. Which of the following structures is not a feature of the mitotic chromosome?
    a. compacted 30 nm chromatin fiber
    b. topoisomerase II
    c. centrosome
    d. kinetochore

12. During which of the following mitotic phases are chromosomes dispersed the most?
    a. prophase
    b. prometaphase
    c. metaphase
    d. anaphase

13. Which type of spindle microtubules are attached to chromosomes?
    a. astral microtubules
    b. kinetochore microtubules
    c. polar microtubules
    d. All of the above spindle microtubules are attached to chromosomes.

14. What is the best evidence that anaphase chromosomes are not pulled to the poles like fish on a line?
    a. Dyneins and kinesin-related proteins are associated with kinetochores.
    b. Depolymerization of spindle microtubules occurs at the kinetochores.
    c. Microtubule disassembly establishes the rate of chromosomal movement.
    d. In fact, it is believed that chromosomes are pulled in like fish on a line.

15. Which of the following experimental observations (some real, some contrived) would not be consistent with our current understanding of cytokinesis?
    a. Anti-myosin antibodies injected in a dividing cell cause cytokinesis to stop.
    b. When the myosin II gene is mutated, cells undergo nuclear division, but not cytokinesis.
    c. Anti-actin antibodies injected into dividing cells have no effect on cytokinesis.
    d. Actin filaments seen in micrographs are aligned parallel to the cleavage furrow.

16. The cleavage furrow is to animal cells as the ___?___ is to plant cells.
    a. cell plate
    b. cell membrane
    c. cell wall
    d. microtubule

17. The life cycle of mosses is characterized by an "alternation of generations," where the small, haploid gametophyte produces gametes, which fertilize and give rise to the tall sporophyte generation. What type of life cycle is this?
    a. gametic
    b. sporic
    c. zygotic
    d. asexual

18. If diplotene lampbrush chromosomes are incubated with radiolabeled uracil and then examined using autoradiography, most of the radiolabel will appear
    a. on the lampbrush loops.
    b. along the lampbrush backbone.
    c. on both the lampbrush loops and the lampbrush backbone.
    d. unaffiliated with the chromosomes.

19. Which stage of cell division provides visual evidence of Mendel's law of independent assortment?
    a. mitotic anaphase
    b. mitotic metaphase
    c. meiotic anaphase I
    d. meiotic anaphase II

20. The following diagram represents a recombination event between homologous chromosomes (one homologue in bold and the other in fine). Did branch migration occur in this crossover, and if so, in which direction?

   a. No branch migration occurred.
   b. Yes, branch migration occurred from right to left.
   c. Yes, branch migration occurred from left to right.
   d. It cannot be determined from the information given.

Problems and Essays

1. You are interested in studying the cytosolic factors that control the phases of the cell cycle. You have set up two fusion experiments. In the first experiment, you fuse G_1 cells with S cells. In the second, you fuse G_2 cells with S cells. Both hybrids are give ^{3}H-thymidine, and then the nuclei from the G_2 and G_1 cells are examined by autoradiography. Using the diagrams below, indicate where radioactivity is found.

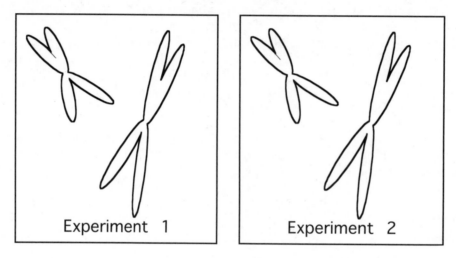

2. Your lab is searching for human genes that code for proteins playing a role in regulating the cell cycle. Because many of the cell cycle proteins are highly conserved, human genes can produce proteins that function perfectly well in yeast cells. You have many resources with which to study this phenomenon, including a culture of human cells, several yeast cell cultures (including some temperature-sensitive mutants that prevent steps of the cell cycle at nonpermissive temperatures), restriction enzymes, and techniques for introducing foreign DNA into yeast cells. Design an experiment to identify and isolate a human gene encoding a cell cycle control protein.

3. The figure below represents the activity curve of MPF isolated from the cytoplasm of frog embryonic cells. Superimpose on this figure a curve representing the concentration of cyclins in the cytoplasm. Now add a curve representing the activity of MPF in the presence of RNase.

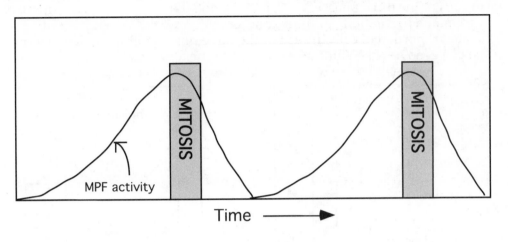

4. The yeast *Schizosaccharomyces pombe* normally divides when it reaches a certain size, as shown in the wild type diagram below. You have isolated two mutants, *cdc25⁻* and *wee1⁻*, but your professor forgot to label which culture was which. This does not pose a problem. Using the microscope, you take a quick look at cells from each of the two cultures of mutants, and determine immediately which culture contains which mutant. Here is what you see:

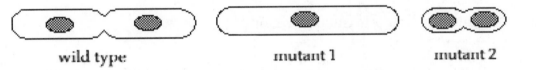

    a. Which mutant, 1 or 2, corresponds to the *cdc25⁻* strain, and which corresponds to the *wee1⁻* strain?

    b. Which phase of the cell cycle is altered by these mutations? Is the phase increased or decreased in mutants 1 and 2?

5. The p53 protein is a transcription factor that initiates production of a cyclin-dependent kinase inhibitor (among other things) when there is damage to DNA. Thus, cells are arrested in G_1 until the damage is repaired. A second checkpoint, at the end of G_2, acts independently of p53. The first graph on the left shows how p53 influences the percentage of cells in each of the three phases, G_1, S, and G_2, both before and 8 hours after DNA-damaging gamma irradiation. In the second graph, draw the bars representing the approximate proportion of cells in each of the three phases following irradiation– phases in which the p53 gene has been mutated and is nonfunctional.

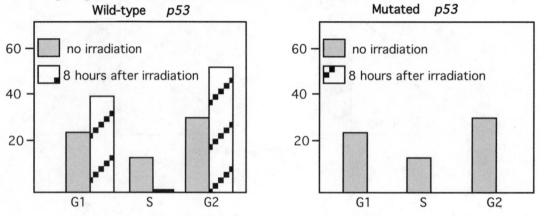

6. For each of the following phases of the cell cycle, indicate whether the chromosome number is haploid (H), diploid (D), or could be either (E). Also indicate whether chromosomes are replicated ® or unreplicated (U).

Chromosome Number   Chromosome Status

_____   _____   a. G_2.

_____   _____   b. meiotic metaphase

_____   _____   c. mitotic prometaphase

_____   _____   d. meiotic pachytene

_____   _____   e. meiotic interkinesis

_____   _____   f. mitotic cytokinesis

_____   _____   g. G_o

7. When genetic recombination occurs in circular chromosomes, as it does in prokaryotic cells or in plasmids, the sorts of intermediate structures that result differ from those in linear molecules.

a. Imagine two circular chromosomes undergoing recombination. Draw the Holliday structure that results.

enlarge this area

b. When the Holliday structure is resolved, the result may be a dimeric circle. Illustrate how this could give rise to two monomeric circles again, both with and without genetic recombination:

Web Explorations

In Chapter 14, the topics of DNA replication and the distribution of chromosomes during cell division was limited to the nuclear genetic material. Recall, however, that mitochondria and chloroplasts have DNA as well. Do the same rules apply to these organelle genomes?

Using the World Wide Web, learn how DNA replication in organelles is similar to and different from that in the nucleus. Are organelle chromosomes as carefully distributed to "daughter" organelles when the organelle divides as nuclear chromosomes are to daughter cells in mitosis and meiosis?

15

Cell Signaling: Communication Between Cells and Their Environment

Learning Objectives

When you have finished this chapter, you should be able to:

1. Identify the steps common to signal transduction pathways.
2. Define first messenger, second messenger, and effector.
   a. Give some examples of first messengers and effectors.
   b. Name some of the second messengers that have been identified to date.
3. Describe specific ways in which second messengers and effectors influence cellular activity.
4. Distinguish between heterotrimeric G proteins and receptor tyrosine kinases.
   a. Compare signal cascades that alter metabolism with those that affect cell growth and division. (Name specific components of both types of pathways.)
   b. Discuss ways in which mutations in genes coding for signal pathway proteins can lead to cancer and other disorders.
5. Give some examples of convergence, divergence, and crosstalk between different signaling pathways.
6. Cite the major experimental findings that have contributed to our understanding of signal transduction.

Key Terms and Phrases

cell signaling	ligand	signal transduction
signal transduction pathways	G protein	cyclic AMP (cAMP)
second messenger	adenylyl cyclase	effector
reaction cascade	protein kinase A (PKA)	heterotrimeric G protein
G protein-coupled receptors	phospholipase C	protein kinase C
calcium-binding proteins	calmodulin	protein tyrosine kinase
receptor tyrosine kinase (RTK)	phosphatidylinositol 3-hydroxy kinase (PI(3)K)	
insulin receptor substrates (IRSs)	SH2 domains	ras oncogene
apoptosis	programmed cell death	

Reviewing the Chapter

I. Introduction
  A. Cells must respond appropriately to external stimuli to survive.
  B. Cells respond to stimuli via cell signaling, which includes:
    1. recognition of the stimulus by a specific plasma membrane receptor.
    2. transfer of a signal across the plasma membrane.

3. transmission of the signal to effector molecules within the cell, which causes a change in cellular activities.
4. cessation of the cellular response due to inactivation of the signal molecule.
C. Some signal molecules enter cells; others bind to cell-surface receptors.

II. Basic Elements of Cell Signaling Systems
A. Extracellular messenger molecules (ligands) transmit messages between cells.
B. Receptors on or in target cells receive the message.
  1. Some cell surface receptors generate an intracellular second messenger.
  2. Other surface receptors recruit proteins to their intracellular domains.
C. Signaling pathways consist of a series of proteins.
  1. Each protein in a pathway alters the conformation of the next protein.
  2. Protein conformation is usually altered by phosphorylation.
    a. Signal transduction pathways have protein kinases and phosphatases.
    b. Phosphorylation stimulates or inhibits the activity of the protein.
  3. Target proteins ultimately receive a message to alter cell activity.
D. Signal transduction is terminated in several ways.

III. A Survey of Extracellular Messengers and Their Receptors
A. Extracellular messengers include:
  1. Small molecules, such as amino acids and their derivatives.
  2. Steroids
  3. Eicosanoids, which are lipids derived from fatty acids.
  4. Various peptides and proteins
B. Receptor types include:
  1. G protein-coupled receptors (GPCRs)
  2. Receptor protein-tyrosine kinases (RTKs)
  3. Ligand gated channels
  4. Steroid hormone receptors
  5. Specific types of miscellaneous receptors, such as B- and T-cell receptors

IV. G Protein-Coupled Receptors and Their Second Messengers
A. G protein coupled receptors (GPCRs) are widespread and numerous.
B. GPCRs have seven transmembrane domains and interact with G proteins.
C. Signal transduction by G protein-coupled receptors:
  1. Ligand binding on the extracellular domain changes the intracellular domain.
  2. Affinity for G proteins increases, and the receptor binds a G protein intracellularly.
  3. GDP is exchanged for GTP on the G protein, thus activating the G protein.
  4. One ligand-bound receptor can activate many G proteins.
D. G proteins
  1. Heterotrimeric G proteins have three subunits: α, β, and γ.
  2. Ligand binding to the receptor replaces GDP with GTP on the α subunit.
  3. G_α leaves the complex and activates an effector, such as adenylyl cyclase.
  4. G_α hydrolyzes the bound GTP to GDP, thus inactiving the response.
  5. There are several forms of the G_α subunit, each of which activates a different receptor.
  6. The $G_{\beta\gamma}$ subunits can also have signaling functions.
E. Termination of the response: Desensitization
  1. Activated GPCRs are phosphorylated by G protein-coupled receptor kinase (GRK).
  2. Arrestins are proteins that compete with G proteins to bind GRCRs.
  3. Arrestins stimulated uptake of phosphorylated GPCRs by clathrin-coated pits.
  4. GPKs are also subject to regulation by phosphorylyation.

F. The Human Perspective: Disorders associated with G protein-coupled receptors:
    1. Several disorders are caused by defects in receptors or G proteins.
    2. Loss of function mutations result in nonfunctional signal pathways.
        a. Retinitis pigmentosa (RP), progressive degeneration of the retina, can be caused by mutations in rhodopsin, some of which involve its ability to activate a G protein.
    3. Gain of function mutations may create a perpetually actived G protein.
        a. Certain benign thyroid tumors (adenomas) are caused by a mutation in a receptor. As a result, the G protein is constantly activated.
        b. Temperature-sensitive mutations of G proteins are implicated in precocious puberty and hypoparathyroidism.
    4. Certain polymorphisms in G protein-related genes may result in an increased susceptibility to asthma, high blood pressure, and certain behaviors, as well as decreased susceptibility to HIV.
G. Bacterial Toxins, such as cholera toxin and pertussis virulence factor, target GPCRs and G proteins.
H. The discovery of a second messenger: Cyclic AMP:
    1. Second messengers are released in the cytoplasm after binding of a ligand.
        a. Cyclic AMP (cAMP) is one type of second messenger.
        b. Second messengers amplify the response to a single extracellular ligand.
    2. Glucose mobilization– an example of a response induced by cAMP:
        a. Cyclic AMP is formed by adenylyl cyclase, an effector.
        b. Binding of a ligand activates adenylyl cyclase.
        c. Cyclic AMP triggers a reaction cascade which amplifies the message.
        d. The cascade starts with the binding of cAMP to cAMP-dependent protein kinase A (PKA).
        e. PKA inhibits glycogen synthase and activates phosphorylase kinase.
I. Lipid-derived second messengers:
    1. Some phospholipids of cell membranes are converted into second messengers by activated phospholipases.
    2. Phosphatidylinositol-Derived Second Messengers:
        a. Phosphatidylinositol-specific phospholipase C-β produces second messengers derived from phosphatidylinositol– inositol triphosphate (IP$_3$) and diacylglycerol (DAG).
        b. DAG activates protein kinase C, which phosphorylates serine and threonine residues on target proteins.
        c. One IP$_3$ receptor is a calcium channel located at the surface of the smooth endoplasmic reticulum. Binding of IP$_3$ opens the channel and allows Ca^{2+} ions to diffuse out.
J. The specificity of G protein-coupled responses:
    1. Both GPCRs and G proteins exist as different isoforms.
    2. The specifity of a cellular response mediated by a GPCR depends on the isoforms involved.
    3. The same stimulus can be excitatory for one GPCR and inhibitory for another.
K. Regulation of blood glucose levels:
    1. Different stimuli acting on the same target cell may induce the same response.
        a. Glucagon and epinephrine bind to different receptors on the same cell.
        b. Both hormones stimulate glucose breakdown and inhibit its synthesis.
        c. cAMP is activated by both hormone receptors' G protein.
    2. Responses are amplified by signal cascades.
    3. Other aspects of cAMP signal transduction pathways:

                a. Some PKA molecules phosphorylate nuclear proteins.
                b. Phosphorylated transcription factors regulate gene expression.
                c. Phosphatases halt the reaction cascade by removing phosphate groups.
                d. Regulation can occur at any stage of the cascade
                e. cAMP is produced only as long as the external stimulus is present.
        L. The role of GPCRs in sensory perception:
            1. Rhodopsin is a photosensitive protein for black-and-white vision that is also a GPCR.
            2. Several color receptors are GPCRs.
            3. Odorant receptors in the nose are GPCRs.
            4. Taste receptors for bitter and some sweet flavors are GPCRs.

V. Protein-Tyrosine Phosphorylation as a Mechanism for Signal Transduction

    A. Protein-tyrosine kinases phosphorylate tyrosine residues on target proteins.
    B. Protein-tyrosine kinases regulate cell growth, division, differentiation, survival, and migration.
    C. Receptor tyrosine kinases (RTKs) are cell surface receptors of the protein-tyrosine kinase family.
    D. The mechanism of action of RTKs:
        1. Receptor dimerization results from ligand binding.
        2. Protein kinase activity is activated.
                a. Tyrosine kinase phosphorylates another subunit of the receptor (autophosphorylation).
                b. RTKs phosphorylate tyrosines within phosphotyrosine motifs.
        3. Phosphorylated tyrosines bind effector proteins that have SH2 domains and PTB domains.
        4. SH2 and PTB domain proteins include:
                a. adaptor proteins that bring in other proteins.
                b. docking proteins that bring in other tyrosine phosphorylation sites on other proteins.
                c. Transcription factors that form dimmers and move to the nucleus to activate transcription of specific genes.
                d. Signaling enzymes (mostly other kinases) that lead to overall biochemical changes in the cell.
        5. The RTK response ends when the RTK is internalized into the cell.
    E. The Ras-MAP kinase pathway:
        1. Ras is a monomeric G protein embedded in the plasma membrane by a lipid group.
        2. Ras is active when bound to GTP and inactive bound to GDP.
        3. Accessory proteins play a role:
                a. GTPase activating proteins (GAPs) shorten the active time of Ras.
                b. Guanine nucleotide-exchange factors (GEFs) stimulate the exchange of GDP for GTP.
                c. Guanine nucleotide-dissociation inhibitors (GDIs) inhibit release of GDP.
        4. Ras activates a cascade of enzymes (the MAP Kinase pathway) resulting in activated transcription factors.
        5. Adapting the MAP kinase to transmit different types of information:
                a. The end result of the MAP kinase cascade differs in different cells and different situations.
                b. Specificity of the MAP kinase response is due to differences in the types of kinases that participate in the cascade.
                c. Specificity is also due to differences in the spatial organization of the components of the cascades.

F. Signaling by the insulin receptor:
    1. Insulin regulates blood glucose levels by increasing cellular uptake of glucose.
    2. The insulin receptor is a protein-tyrosine kinase.
        a. Autophosphorylated receptor associates with insulin receptor substrate proteins (IRSs).
        b. IRSs bind proteins with SH2 domains, which activate downstream signal molecules.
        c. One SH2 domain-containing protein is a kinase that phosphorlyates a lipid, PI3-kinase.
    3. PI3 kinase activates proteins that regulate glucose uptake by GLUT4 transporters.
        a. GLUT4 transporters reside in intracellular membrane vesicles.
        b. Vesicles fuse with the plasma membrane in response to ligand binding to the insulin receptor.
    4. Diabetes mellitus is caused by defects in insulin signaling.
    5. Type II diabetes is caused by gradual insensitivity to insulin.
G. Signaling pathways in plants:
    1. Plants lack cyclic nucleotides and RTKs.
    2. Plants have protein kinases that phosphorylate histidine residues.
        a. The downstream cascade is similar to MAP kinase cascade.
        b. The target of the cascade is usually transcription factors.

VI. The role of calcium as a second messenger:

A. Cytoplasmic calcium levels are determined by events within a membrane.
    1. Calcium levels are low in the cytosol because it is pumped out into the extracellular space and the membrane is highly impermeable to the ion.
    2. Calcium channels can be transiently opened by action potential (in muscle cells) or calcium itself.
    3. Calcium binds to calcium-binding proteins (such as calmodulin), which affects other proteins.
B. Regulating calcium concentrations in plant cells:
    1. Cytosolic calcium changes in response to several stimuli, including light, pressure, gravity, and hormones.
    2. Calcium signaling aids in decreasing turgor pressure in guard cells.

VII. Convergence, Divergence, and Crosstalk Among Different Signaling Pathways

A. Information circuits within cells have a number of common features.
    1. Signals from unrelated receptors can converge to activate a common effector.
    2. Identical signals can diverge to activate a variety of effectors.
    3. Signals can be passed back and forth between pathways via crosstalk.
B. G protein coupled receptors, receptor tyrosine kinases, and integrins bind to different ligands but they all can lead to a docking site for Grb2.
C. Binding of EGF or PDGF to its receptor can activate effectors of several divergent pathways, such as Grb2 (MAP kinase cascade), PI(3)K, and phospholipase C.
D. Crosstalk between signaling pathways:
    1. Cyclic AMP can block signals transmitted through the MAP kinase cascade.
    2. Ca^{2+} and cAMP can influence each other's pathways.

VIII. The Role of NO as an Intercellular messenger

A. Nitric oxide (NO) is both an intercellular and intracellular messenger with a variety of functions.
B. NO is produced by nitric oxide synthase.
    1. NO stimulates guanylyl cyclase, making cGMP.
    2. cGMP decreases cytosolic calcium and relaxes smooth muscle.
    3. NO also plays a role in male arousal.

    a. Viagra works by inhibiting breakdown of cGMP.
    b. cGMP retains smooth muscle relaxation and erections.
IX. Apotosis (Programmed Cell Death)
    A. Cells that are no longer needed or can harm the individual undergo apoptosis,
        programmed cell death.
        1. Apoptosis is an ordered process involving cell shrinkage, loss of adhesion to
            other cells, dissection of chromatin, and engulfment by phagocytosis.
        2. These changes are activated by proteolytic enzymes called caspases, which
            target:
            a. Protein kinases, some of which cause detachment of cells.
            b. Lamins, which line the nuclear envelope.
            c. Cytoskeletal proteins.
            d. CAD, or caspase activated Dnase.
    B. The extrinsic pathway of apoptosis is initiated by external stimuli:
        1. TNF is detected by a TNF cell surface receptor.
        2. Activated (bound) TNF receptors recruit procaspases to the intracellular
            domain of the receptor.
        3. Procaspases convert other procaspases to caspases
        4. Caspases activate executioner caspases, leading to apoptosis.
    C. The intrinsic pathway of apoptosis is initiated by intracellular stimuli.
        1. Proapoptotic proteins of the Bcl-2 family stimulate mitochondria to leak
            proteins.
            a. Mitochondrial proteins and procaspases form a complex that initiates
                cell shrinkage and membrane lipid scrambling.
            b. Inner leaflet membrane lipids that appear on the outer leaflet signal
                phagocytotic cells to eat the apoptotic cell.
            c. Cell contents are not spilled, preventing inflammation.
        2. Antiapoptotic proteins promote survival.
        3. Cell fate depends on the balance between pro- and antiapoptotic signals.
X. Experimental Pathways on the Web: The Discovery and Characterization of GTP-Binding
    Proteins (G Proteins)
    A. Certain hormones, when bound to a receptor, activate adenylyl cyclase on the inner
        side of the membrane.
        1. Various receptors stimulate a common population of adenylyl cyclases.
        2. Receptors and adenylyl cyclase effectors exist separately.
    B. Guanyl nucleotides regulate the response of adenylyl cyclase to the hormone.
        1. GTP changes the affinity of hormone-binding sites for their hormones.
        2. GTP stimulates adenylyl cyclase activity.
        3. Binding of GTP, but not its hydrolysis, stimulates cAMP formation.
        4. Hydrolysis of GTP inactivates adenylyl cyclase.
        5. Activation of the receptor leads to the release of bound GDP and binding of
            GTP.
    C. There is a regulatory component between the hormone receptor and the adenylyl
        cyclase effector. This component is inactivated by cholera toxin.
    D. GTP-binding protein consists of three subunits.
    E. The α subunit of the G protein has the GTP-binding site and activates adenylyl
        cyclase.

Key Figure

Figure 15.2 Signal transduction pathways consist primarily of protein kinases and protein phosphatases whose catalytic actions change the conformation, and thus the activities, of the proteins they modify.

Questions for Thought:

1. This generalized signal transduction pathway requires several different proteins, including protein kinases and protein phosphatases. What advantage might this multistep pathway provide for the cell that a simpler and more direct pathway does not?
2. This same generalized signal transduction cascade occurs in many different cells and affects many different changes in cellular activity. Given the similarities in these proteins and their activities, how can you account for the differences in effects?
3. Likewise, many different stimuli may activate this same pathway in different cells. How can you account for the same effects given the differences in stimuli?
4. How might different mutations in the genes that encode these proteins influence their function?

Review Problems

Short Answer

1. Give the significance (not the definition) of the following terms or phrases. Say what they do or why they are important. For example:

   G proteins: *carry the signal generated by ligand binding at the receptor to an effector molecule.*

   a. the many steps of a signal transduction cascade

   b. cAMP

   c. phosphatases

   d. calmodulin

   e. SH2 domain

   f. receptor tyrosine kinases

   g. Ras protein

   h. MAP kinase cascade

   i. crosstalk

   j. apoptosis

2.   Compare and contrast the following:
   a.   kinase vs. phosphatase

   b.   heterotrimeric G protein vs. monomeric G protein

   c.   phospholipase C vs. protein kinase C

   d.   tyrosine kinases vs. serine/threonine kinases

   e.   signaling pathways in plants vs. signaling pathways in animals

Multiple Choice

1.   Cell signaling means all of the following except:
        a. intercellular communication.
        b. export of cellular waste products.
        c. environmental monitoring.
        d. response to stimuli.

2.   In a signal transduction pathway, what is passed from the upstream member of the pathway to the downstream member?
        a. electrons
        b. protons
        c. phosphates
        d. information in the form of alterations in protein conformation

3.   Both epinephrine and glucagon bind to G protein-coupled receptors and activate glycogen breakdown. Therefore, epinephrine and glucagon must:
        a. have very similar structures and bind to the same receptor.
        b. bind to receptors with different ligand-binding sites but similar functions.
        c. bind to different receptors and activate different second messengers.
        d. bind to the same receptors, one being intracellular and the other, extracellular.

4.   If you break liver cells by homogenization, separate the broken cell membranes (particulate) from the cytoplasm (soluble), and add epinephrine to the soluble portion, the result will be:
        a. the production of cAMP.
        b. binding of the epinephrine to its receptor.
        c. activation of adenylyl cyclase.
        d. none of the above.

5. Glucagon binds to the glucagon receptor, which interacts with a membrane-bound G protein, whose α subunit releases GDP and binds GTP. The α subunit then dissociates from the G protein and diffuses to a membrane-bound adenylyl cyclase which produces cAMP. In this scenario, name the (1) transducer, (2) effector, (3) first messenger, and (4) second messenger.
    a. (1) glucagon (2) cAMP (3) G protein (4) GTP
    b. (1) G protein (2) adenylyl cyclase (3) glucagon (4) cAMP
    c. (1) G protein (2) GTP (3) glucagon (4) cAMP
    d. (1) glucagon (2) G protein (3) adenylyl cyclase (4) cAMP

6. Epinephrine binds to the same type of receptors in liver, fat, and smooth muscle cells. Yet in liver, glycogen breaks down; in fat, triacylglycerols break down; and smooth muscle cells relax. How can the same hormone produce three such different responses?
    a. The same receptors activate three different second messengers.
    b. The same second messengers activate three different intracellular cascades.
    c. The same enzyme that breaks down glycogen also breaks down fat and relaxes muscle.
    d. None of the above.

7. G proteins are said to be self-inactivating. What G-protein function accounts for this?
    a. GTPase activity of the α subunit
    b. activation of the effector
    c. binding to the receptor
    d. dissociation of the subunits

8. People who suffer from thyroid adenomas have constitutively activated G proteins in some cells of the thyroid gland. Tumor cells not only secrete copious quantities of thyroid hormones but also divide and grow excessively. Based on these observations, in what kind(s) of signal transduction pathways do the mutant G proteins participate?
    a. cAMP-regulated pathways
    b. MAP kinase pathways
    c. both cAMP and MAP kinase pathways
    d. neither cAMP nor MAP kinase pathways

9. What factors are responsible for the specificity of G protein-coupled responses?
    a. Protein kinases can phosphorylate different proteins in different cells.
    b. Receptors can exist as isoforms.
    c. G-protein subunits can exist as isoforms.
    d. All of the above.

10. Acetylcholine on a smooth muscle cell is to phospholipase C as epinephrine on a liver cell is to __?__.
    a. protein kinase C
    b. cAMP
    c. adenylyl cyclase
    d. GTP

11. Serine and threonine are to __?__ as tyrosine is to receptor tyrosine kinases.
    a. phosphatases
    b. G proteins
    c. cAMP
    d. protein kinases A and C

12. Ryanodine and IP$_3$ receptors are found on __?__ membranes, whereas epinephrine and glucagon receptors are found on __?__ membranes.
    a. plasma, intracellular
    b. intracellular, plasma
    c. mitochondrial, endoplasmic reticulum
    d. outer, inner

13. Depending on the cell, increases in intracellular Ca^{2+} can do all of the following except:
    a. stimulate the increase of intracellular calcium.
    b. stimulate the removal of intracellular calcium.
    c. stimulate the breakdown of cAMP.
    d. stimulate the breakdown of calmodulin.

14. The sequence of events that occurs after insulin binds to its receptor tyrosine kinase is:
    a. binding of IRS —> phosphorylation of proteins with SH2 domains —> effect
    b. binding of proteins with SH2 domains —> phosphorylation of IRS —> effect
    c. autophosphorylation and phosphorylation of IRS —> binding of proteins with SH2 domains —> effect
    d. autophosphorylation and binding of IRS —> phosphorylation of proteins with SH2 domains —> effect

15. Proteins with SH2 domains are to __?__ as adenylyl cyclase and phospholipase C are to G protein-coupled receptors.
    a. IRSs
    b. receptor tyrosine kinases
    c. insulin
    d. autophosphorylation

16. Which mutant form of ras is likely to cause malignancy?
    a. ras that cannot hydrolyze GTP
    b. ras that cannot bind to GTP
    c. ras that cannot bind to Grb2 or Sos
    d. ras that cannot bind to Raf

17. The MAP kinase cascade stimulates transcription via specific transcription factors, and stimulates translation via
    a. SREs
    b. MAPK
    c. PHAS-I and eIF4E
    d. all of the above

18. What second messenger has been implicated in blood pressure regulation?
    a. cAMP
    b. NO
    c. CO
    d. IP3

19. A mutation in the FAK gene would likely result in:
    a. abnormal cell shape.
    b. abnormal wound healing.
    c. abnormal cell growth.
    d. all of the above.

20. A cell responds to an external stimulus by increasing the intracellular Ca^{2+} concentration, releasing IP_3 from the membrane, and producing salicylic acid. What kind of cell does this describe?
   a. bacterial
   b. viral
   c. plant
   d. animal

Problems and Essays

1. In your laboratory, you have a culture of human cells that express the receptors for both platelet-derived growth factor (PDGF) and epidermal growth factor (EGF). You can add either PDGF or EGF to your culture and the cells enter the S phase and begin DNA replication (treatment a). You have done a series of experimental manipulations (treatments b through f), and tested for DNA replication after each one. Write your results in the third column of the table below.

Treatment	Additions	DNA Replication
a. none	PDGF or EGF	yes
b. microinjection (intracellular) of constitutively active Ras protein	none	
c. microinjection of saline ("mock" injection)	PDGF or EGF	
d. microinjection of saline ("mock" injection)	none	
e. microinjection of anti-Ras antibodies	PDGF or EGF	
f. microinjection of anti-Ras antibodies	none	

2. The nicotinic acetylcholine receptor is membrane-bound protein that also functions as a ligand-gated ion channel. Upon binding acetylcholine, the channel opens and sodium ions pass into the cell, flowing in the direction of their electrochemical gradient. Put a check next to all of the features listed below that apply to the nicotinic acetylcholine receptor.

   a._____Ligand binding to the receptor is reversible.
   b._____Low concentrations of GTP stimulate the dissociation of the ligand from the receptor.
   c._____Ligand binding results in increased intracellular concentrations of cAMP.
   d._____Ligand binding changes the voltage measured across the cell membrane.
   e._____The effects of ligand binding cannot be reconstructed in homogenized cells, even if both soluble and particulate fractions are included.
   f._____Mutations in the ras gene will influence the effect of ligand binding.

3. A single mutation in the ras gene (resulting in the substitution of a valine for glycine at amino acid position 12 of the protein) reduces the GTPase activity of the protein and transforms the cell from normal to malignant. Explain.

4. When one molecule of epinephrine binds a ß-adrenergic receptor, 500,000 protein kinase A molecules are activated per second. One stage of the amplification occurs at the level of adenylyl cyclase, which can produce 1000 cAMPs per second. You are interested in finding where the other 500-fold amplification occurs, so you perform the following experiment: Using radioactive GppNp, a GTP analog that cannot be hydrolyzed, you determine that when 0.8% of a mole of receptors are bound to epinephrine, four moles of GppNp are bound to G proteins. Based on your data, where does the additional 500-fold amplification occur?

5. Serotonin is a small molecular weight amine that can act as a neurotransmitter, conveying a signal between adjacent nerve cells, or as a hormone, entering the blood stream and conveying a signal to cells in nonadjacent tissues. It has been implicated in personality, mood, and sleep, and also in blocking pain at the level of the central nervous system. You are interested in establishing the mechanism by which serotonin affects target cells. Your lab has made the following observations:
    i.   Serotonin causes an increase in cAMP in target cells.
    ii. The increase in cAMP can be observed in suspensions of homogenized cells, but not if the particulate fraction has been removed by centrifugation.
    iii. Dissociation of serotonin from membrane fragments in homogenized cells requires the presence of GTP.
    iv. Membranes of target cells have GTPase activity. When target cells that are sensitive to both serotonin and epinephrine are given both hormones, the total effect is not additive.
    Propose a mechanism of action for serotonin.

Web Explorations

Opiates and opium-like substances, including heroin and morphine, act as painkillers and euphoriates by binding to G protein-coupled receptors in the central nervous system. These substances, first isolated from the poppy plant, have been used both medicinally and for recreation for over 5000 years. The unfortunate side effect is addiction. Although the exact mechanism of addiction is only partly understood, it clearly involves molecular changes at the level of the cell signaling pathway invoked by opiate receptor binding.

Use the World Wide Web to learn more about the mechanism of drug addiction from the perspective of a cell signaling pathway. What components of the pathway are altered with prolonged or repeated exposure to opiates? Are the changes reversible? How is our understanding of the molecular characteristics of addiction contributing to new therapies for recovering addicts?

16

Cancer

Learning Objectives

When you have finished this chapter, you should be able to:

1. Define cancer. Discuss the characteristics of benign tumors, malignant tumors, and secondary tumors (metastases).
2. Describe the phenotypic characteristics of transformed cells and how these characteristics differ from those of normal cells.
3. Give an overview of known causes of cancer.
4. Describe the initiation and the promotion phases of tumorogenesis. Explain tumor progression on the basis of cancer genetics.
5. Distinguish between tumor suppressor genes and oncogenes.
    a. Describe how mutations in these genes can affect cellular activity.
    b. Explain how mutations in these two types of genes cause respective differences in the heritability of cancers.
    c. Give specific examples of these two types of genes.
    d. Describe some cancers, including their underlying cellular causes, that have been attributed to these genes.
6. Discuss the relationship of DNA and RNA tumor viruses to cellular genes; recognize the role of these viruses in advancing our understanding of cancer.

Key Terms and Phrases

cancer	malignant tumor	benign tumor
secondary tumor (metastasis)	transformed cells	tumor-associated antigens
anchorage dependence	DNA tumor viruses	RNA tumor viruses
monoclonal	tumorigenesis	initiation
promotion	tumor-suppressor genes	RB gene
p53 protein	oncogenes	proto-oncogenes

Reviewing the Chapter

I. Introduction

    A. Cancer results from alterations in the DNA of a somatic cell during the lifetime of the affected individual.
        1. Gene expression in cancer cells is altered.
        2. Cancer cells proliferate in an uncontrolled manner, forming malignant tumors.
        3. Malignant tumors tend to metastasize (establish secondary tumors).
    B. Cancer is the focus of a massive research effort.

1. Current treatments for cancer simultaneously kill cancer cells and normal cells.

II. Basic Properties of a Cancer Cell

A. Malignant cells are not responsive to influences that cause normal cells to cease growth and division.
1. The capacity for growth and division is similar between cancer cells and normal cells.
2. When there are no growth factors in the medium or when cells contact surrounding cells:
   a. Normal cells stop growing.
   b. Malignant cells continue to grow.
B. The phenotype of a cancer cell:
1. In culture, normal cells can be transformed by chemicals or viruses.
2. Different types of cancer cells share a number of similarities.
   a. Cancer cells often have aberrant chromosome numbers (aneuploidy).
   b. The cytoskeleton of cancer cells is disorganized and reduced.
   c. Cancer cells possess certain cell surface proteins called tumor-associated antigens.
   d. Cancer cells are less adhesive.
   e. Cancer cells are motile.
   f. Cancer cells ignore signals transmitted by neighboring cells.
   g. Cancer cells do not need growth factors for proliferation.
   h. Cancer cells do not need a solid surface for growth (lack of anchorage dependence).
   i. Cancer cells divide indefinitely, a characteristic often attributed to the presence of telomerase.

III. The Causes of Cancer

A. Mutagenic agents, such as carcinogenic chemicals and radiation can cause cancer by altering the genome.
B. DNA tumor viruses and RNA tumor viruses carry genes whose products interfere with cell growth regulation.
C. Causes of most types of human cancer are still unknown.
D. Diet can influence the risk of developing cancer.

IV. The Genetics of Cancer

A. The development of a malignant tumor is a multistep process.
1. Cancer results from the uncontrolled proliferation of a single cell.
2. Tumorigenesis occurs by a cumulative progression of genetic alterations.
   a. During tumorigenesis, cells become less responsive to growth regulation and better able to invade normal tissues.
   b. The first step is the formation of a benign tumor which is a tumor composed of cells that proliferate uncontrolled but cannot metastasize to other sites.
   c. The products of the genes involved in carcinogenesis are usually responsible for cell cycle regulation, cell adhesion, and DNA damage repair.
   d. The sequence in which genes mutate influences the development of cancer.
3. DNA microarrays (DNA chips) are used to diagnose cancer.
   a. DNA corresponding to thousands of different genes is spotted on a glass plate such that each spot corresponds to a different gene.
   b. Fluorescent cDNA is prepared from cancer cells and allowed to hybridize with the spots.
   c. Spots corresponding to actively transcribed genes fluoresce.

　　　　　d. Different tumors have characteristic transcription patterns.
　　　　　e. DNA arrays may help doctors to design better treatment regimens.
　　　4. Cancers often arise in tissues that are engaged in a high level of cell division.
　B. Tumor-suppressor genes and oncogenes: Brakes and accelerators:
　　　1. Tumor-suppressor genes encode proteins that restrain cell growth.
　　　　　a. A normal cell fused to a cancer cell can suppress malignant characteristics in the latter.
　　　　　b. Specific regions of chromosomes are deleted in cells of certain cancers.
　　　2. Oncogenes encode proteins that promote the loss of growth control and the conversion of a cell to a malignant state.
　　　　　a. Oncogenes were first discovered in genomes of tumor viruses.
　　　　　b. An oncogene is an altered cellular gene (proto-oncogene).
　　　　　c. Proto-oncogenes encode proteins that function in a cell's normal activities.
　　　　　d. Oncogenes act dominantly.
　　　3. For a cell to become malignant, both alleles of a tumor-suppressor gene must be lost, and a proto-oncogene must be converted into oncogene.
　C. Tumor-Suppressor Genes
　　　1. Most of the proteins encoded by tumor-suppressor genes act as negative regulators of cell proliferation.
　　　2. RB gene was the first tumor-suppressor gene to be discovered.
　　　　　a. Retinoblastoma is inherited in certain families, and occurs sporadically in the population at large.
　　　　　b. The cells of children with inherited retinoblastoma have a deletion in one copy of the RB gene.
　　　　　c. The development of retinoblastoma requires both copies of RB to be altered or eliminated.
　　　　　d. People affected with the inherited form are also at high risk of developing other types of tumors later in life.
　　　　　e. The reintroduction of a wild-type RB gene is sufficient to suppress a cell's cancerous phenotype.
　　　3. The role of pRb in regulating the cell cycle:
　　　　　a. The protein encoded by the RB gene (pRb) regulates the passage of cells from G_1 to S phase.
　　　　　b. Transcription factors of E2F family are targeted by pRb.
　　　　　c. The arrest of the cell cycle in G_1, required for normal cell differentiation, is directed by pRb.
　　　　　d. Animals with one mutated copy of the RB gene have an elevated risk of developing cancer.
　　　4. The role of p53: Guardian of the genome:
　　　　　a. The p53 protein suppresses the formation of tumors and maintains genetic stability.
　　　　　　　(1) The p53 gene may be the most important tumor suppressor gene in human genome.
　　　　　　　(2) Fifty percent of human cancers have mutations in p53.
　　　　　　　(3) Tumor cells with p53 mutations are highly invasive.
　　　　　　　(4) Proper functioning of the p53 protein is very sensitive to even slight changes in the amino acid sequence.
　　　　　b. The p53 protein acts as a transcription factor, activating the expression of a gene that inhibits the G_1 -S transition.
　　　　　c. When a cell sustains damage to its DNA, the concentration of p53 rises so that the damage can be repaired before initiating DNA replication.
　　　　　d. The p53 protein triggers apoptosis in cells whose DNA is damaged beyond repair.

5. Other tumor suppressor genes:
    a. Mutations of tumor-suppressor genes that are not RB or p53 are detected in only a few types of cancer.
    b. Colon cancer is often caused by an inherited deletion in a tumor-suppressor gene APC.
    c. Inherited breast cancer is caused by mutations in BRCA tumor suppressor genes, which probably act as transcription factors and as a part of the cell's DNA repair machinery.
D. Oncogenes:
    1. Most of the known oncogenes are derived from proto-oncogenes that are involved in the control of cell growth.
    2. Oncogenes that encode growth factors or their receptors:
        a. Simian sarcoma virus contains the oncogene (sis) which is derived from a cellular gene which codes for the growth factor PDGF.
        b. Oncogene erbB directs the formation of an altered EGF receptor that stimulates the cell regardless of the presence of growth factor.
        c. Some malignant cells contain more surface receptors than normal cells and are therefore sensitive to lower concentrations of growth factors.
    3. Oncogenes that encode cytoplasmic protein kinases:
        a. Raf (a serine/threonine protein kinase in the MAP kinase cascade) can be converted into an oncogene by mutations that turn it into an enzyme that is always "on."
        b. The oncogene product of src is a protein tyrosine kinase which phosphorylates proteins involved in signal transduction, control of cytoskeletal configuration and cell adhesion.
    4. Oncogenes that encode nuclear transcription factors:
        a. A number of oncogenes encode proteins that may act as transcription factors.
        b. Myc protein stimulates cells to reenter cell cycle from G_o stage.
        c. Overexpression of myc may cause cells to proliferate uncontrollably.
    5. Oncogenes that encode products that affect apoptosis:
        a. Alterations that tends to diminish cell's ability to self-destruct increase likelihood of a cell giving rise to a tumor.
        b. The overexpression of blc-2 gene leads to suppression of apoptosis, allowing abnormal cells to proliferate into tumors.
E. The mutator phenotype: Mutant genes involved in DNA repair:
    1. Mismatch repair defects predispose cells to abnormally high mutation rates, which increases the risk of malignancy.
    2. A deficiency in the nucleotide excision repair (NER) system may be involved in a hereditary syndrome, xeroderma pigmentosum.
    3. A deficiency in NER is implicated in hereditary colon cancer.
    4. Tumor cells of this form of cancer often have variations in the length of microsatellite sequences, due to errors during replication.
    5. Some believe that cancer develops from catastrophic phenomena, not gradually.

V. New Strategies for Combating Cancer

A. The widely-used mouse model for testing treatments for cancer may not be as predictive for humans as originally thought.
B. New strategies include using the immune system, introducing a gene that kills tumors cell, inhibiting the activity of cancer-promoting proteins, and preventing the growth of blood vessels that nourish the tumor.
C. Immunotherapy

1. Tumor vaccines utilize a strategy that seeks to stimulate the patient's own immune system to respond to tumor cells.
2. Dendritic cells taken from cancer patients can be made to display tumor proteins, then injected back into the patient.
    a. Dendritic cells are the best of the immune cells for inducing a powerful immune response.
    b. The patient's immune system rejects the tumor.
D. Gene Therapy
    1. Gene therapy modifies a patient's genotype by the addition, deletion, or alteration of a specific gene.
    2. A tumor cell should be converted to the nonmalignant state by acquisition of a wild-type tumor suppressor gene through infection with an engineered virus carrying the gene.
        a. The limitation to this approach is that all tumor cells must be permanently altered.
        b. Adenovirus is being tested to infect, lyse, and destroy tumor cells.
E. Inhibiting the activity of cancer-promoting proteins might stop the uncontrolled growth and invasive properties or malignant cells.
F. Inhibiting the formation of new blood vessels (angiogenesis):
    1. Compounds that inhibit angiogenesis are promising treatments for tumors.
    2. Antibodies to VEGF, a vascular growth factor, are in phase II trials.
    3. Antiangiogenetic compound denies tumors access to the nutrients and oxygen they need to grow.

VI. Experimental Pathways: The Discovery of Oncogenes
A. Experiments of Rous (1911) showed that a tumor could be transmitted from one animal to another by a virus.
B. RNA-dependent DNA polymerase (reverse transcriptase) is the enzyme that replicates the genetic material of RNA tumor viruses.
C. The capacity to transform a cell resides in a restricted portion of the viral genome.
D. The transforming genes of the viral genome (oncogenes) are not true viral genes but are cellular genes previously picked up by RNA tumor viruses.
E. The src gene was the first oncogene identified; it is present in all vertebrate classes.
    1. Both viral and cellular src genes (v-src and c-src, respectively) code for protein kinases.
    2. The viral version of the gene has higher activity than the cellular version.
    3. Increased activity of an oncogene product can transform a normal cell into a malignant cell.
F. Cellular genes can be converted into oncogenes by carcinogenic chemicals or by a mutation in sequences regulating their expression.
G. Single base substitution mutations can convert proto-oncogenes into oncogenes.

Key Figure

Figure 16.11.   Contrasting effects of mutations in tumor-supressor genes and in oncogenes.

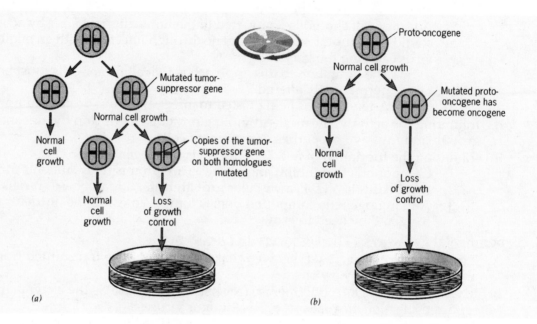

Questions for Thought:

1.   Which type of mutation would be expressed as a genetic dominant and which as a recessive? Why?
2.   The RNA and DNA tumor viruses carry oncogenes that are highly homologous to normal cellular genes. Why wouldn't a virus with a tumor suppressor gene homologue be a tumor virus?
3.   What kinds of cellular functions are carried out by the gene products of normal tumor suppressor genes? The gene products of normal proto-oncogenes?
4.   Assuming you could treat cancers with gene therapy, how would the therapy in cancers involving tumor-suppressor genes differ from that in cancers involving oncogenes?

Review Problems

1. Give the significance (not the definition) of the following terms or phrases. Say what they do or why they are important. For example:
   metastasis: *responsible for the invasive and potentially deadly nature of cancer.*
   a. aneuploidy in transformed cells

   b. Percival Pott's studies of British chimney sweeps

   c. tumor-suppressor genes

   d. proto-oncogenes

   e. chromosomal deletion in children with retinoblastoma

   f. the oncogene, sis, of the simian sarcoma virus

2. Compare and contrast the following:
   a. benign vs. metastatic tumor

   b. tumor virus genes vs. oncogenes

   c. initiators vs. promoters

d. tumors in children with familial retinoblastoma vs. those with sporadic retinoblastoma

e. oncogenes that encode growth factors vs. those that encode growth-factor receptors

Multiple Choice

1. How might you distinguish between a malignant tumor and a benign tumor?
    a. Using microscopy, compare cell shape.
    b. Using microscopy, compare the chromosome integrity.
    c. Using cell culture, compare cell life span.
    d. All of the above.

2. Which of the following proteins has been implicated in aneuploidy of cancer cells?
    a. p53
    b. pRB
    c. Myc
    d. bcl-2

3. All cancer-causing agents identified thus far have what feature in common?
    a. They are all found in cigarette smoke.
    b. They are all mutagenic.
    c. They are all chemicals.
    d. They do not have any features in common.

4. Cancer is said to be monoclonal. What does that mean?
    a. Cancer arises only once in an individual's lifetime.
    b. Cancer stimulates the production of antibodies.
    c. Cancer can only be cured using one treatment.
    d. Cancer results from the uncontrolled proliferation of one wayward cell.

5. In what way is tumor progression like biological evolution?
    a. Survival of cancer victims is like survival of the fittest.
    b. Cancer cells that proliferate fastest become the most abundant in the tumor.
    c. Tumor progression requires both initiation and promotion.
    d. Cells of benign tumors cannot disperse to distant tissues.

6. Of agents that can act as either initiators or promoters, which can be mutagenic?
    a. initiators
    b. promoters
    c. both initiators and promoters
    d. neither initiators nor promoters

7. In some breast cancers, estrogen acts as a(n):
    a. initiator.
    b. promoter.
    c. progressor.
    d. mutagen.

8. Cancers that arise due to mutated tumor suppressor genes require that
    a. one allele of the gene undergoes mutation.
    b. both alleles of the gene undergo mutations.
    c. neither allele of the gene undergoes mutation.
    d. none of the above

9. In children with retinoblastoma, the chromosomal deletion that is present in all of their cells indicates that the chromosomal aberration:
    a. arose somatically in retinal cells.
    b. first occurred in the retina, and then spread to other cells.
    c. arose during early embryonic development.
    d. was inherited from a parent.

10. Would you expect to find that the cells of children with nonfamilial (i.e., sporadic) retinoblastoma have the chromosomal deletion referred to in number (9) above?
    a. yes
    b. no
    c. no way of telling
    d. not enough information given

11. The normal protein product of the RB gene acts as an:
    a. inhibitor of transcription.
    b. activator of transcription.
    c. inhibitor of translation.
    d. activator of replication.

12 Which of the following is not a normal function of p53?
    a. It activates transcription of a cell cycle inhibitor.
    b. It plays a role in DNA repair.
    c. It triggers apoptosis in damaged cells.
    d. It plays an essential role in the embryonic development of the mouse.

13. When a culture of rodent cells is infected with a DNA tumor virus whose gene products bind to p53 and RB, all the cells take on characteristics of malignancy. If the viral gene coding the p53-binding protein is mutated, but that coding for the RB-binding protein is not mutated, how will infected cells respond?
    a. They will become malignant.
    b. They will show no effects of the virus.
    c. They will undergo apoptosis.
    d. They will enter the S phase.

14. The textbook states that cells of patients with adenomatous polyposis coli (APC) "contain a deletion of a small portion of chromosome #5, which was subsequently identified as the site of a tumor-suppressor gene called APC." In which cells of these patients would you expect to find this chromosomal deletion?
    a. all cells
    b. cells of the polyps of the colon epithelium
    c. germ cells
    d. cancer cells

15. When the genes responsible for familial breast cancer were sequenced, it was found that the polypeptides they encoded contained a zinc finger motif. What does this suggest to you about the normal functions of the products of these genes?
    a. The genes code for hormones.
    b. The genes code for growth factors.
    c. The genes code for growth factor receptors.
    d. The genes code for transcription factors.

16. Tumor-suppressor genes usually act as __?__ genes, and oncogenes usually act as __?__ genes.
    a. dominant, dominant
    b. recessive, recessive
    c. dominant, recessive
    d. recessive, dominant

17. Which of the following types of genes have not been identified as oncogenes?
    a. genes that encode growth factor receptors
    b. genes that encode membrane-bound ion channels
    c. genes that encode cytoplasmic protein kinases
    d. genes that encode proteins involved in mismatch repair

18. An oncogene codes for a growth factor receptor that is constitutively active. Of the following statements about a culture of cells expressing this gene, which would you expect to be true?
    a. The cells would be stimulated to divide by addition of growth factors to the medium.
    b. Cells in the culture would have no response to the addition of growth factors.
    c. Cells in the culture would be arrested in the G_0 phase of the cell cycle.
    d. None of the above.

19. The genes responsible for cancer in RNA and DNA tumor viruses are:
    a. unique sequences found only in the viral genomes.
    b. also found in cells of the host, but only after the cells have been transformed.
    c. also found in cells of the host, even when the cells are normal.
    d. also found in many pathogenic bacteria.

20. Of the following statements explaining how alterations of the ras gene cause cancer, which is true?
    a. A base substitution in the c-ras gene that replaces a critical glycine with another amino acid.
    b. Infection of the cell by a tumor virus carrying a v-ras gene.
    c. Linkage of the c-ras gene to a viral promoter.
    d. All of the above statements are true.

Problems and Essays

1. The following have been used as treatments for cancer patients, but none of these approaches has been entirely effective. For each one, briefly explain the theory of how the treatment fights cancer cells, and why the treatment has not cured cancer.

    a. chemotherapy

b. gene therapy

c. surgery

d. stimulating the immune system

e. anti-angiogenic compounds

2. The human population of the world is about six billion (6×10^9). Each person has an average of 100 trillion cells (100×10^{12}) and a life span of 50 years. About one third of the population develops cancer in a lifetime. Calculate the total number of human cells that become malignant per year.

3. You have isolated an RNA tumor virus that is capable of inducing leukemia in mice, but you do not know the mechanism, or the nature of the viral gene product. You also have a mutant strain of the virus that cannot induce murine leukemia and is missing 15% of the genome.

a. Design an experiment to isolate the portion of the normal viral genome that causes the cancer.

b. How might you go about identifying both the gene product of the viral oncogene and making an educated guess about its function?

4. You are testing the growth of both normal and transformed cells, and the effects of the growth factor EGF on both. The following is a graph of the growth curve of normal cells, both with and without added EGF.

a. On the graph, superimpose two curves respectively representing transformed cells both with EGF and without EGF.

b. Now add a curve showing how the addition of EGF on day six would alter the normal, "without EGF" curve.

c. Cells in the normal +EGF and -EGF cultures are in what phase of the cell cycle after day four?

5. Why do tumor-suppressor genes act as recessives, and oncogenes act as dominants?

6. Some cancers develop in two phases, initiation and promotion. Explain at a cellular level why the promoting influence can occur months or even years after initiation and still be effective in promoting cancer?

7. In your lab you have several mouse cell lines growing in culture, designated culture A through culture D. Some of these cell lines are transformed and others are not. For each of the descriptions below, tell whether the cell line is transformed or normal, and if it is transformed, make an educated guess about the gene involved in the transformation.

   a. Cell line A is insensitive to the addition of growth factors to the medium. If the cells are exposed to damaging radiation, the growth curve is unaffected. Many of the cells have abnormal numbers of chromosomes.

   b. Cell line B will not grow suspended in liquid media. The cells require a hard substrate, and then they only grow until one cell layer covers the substrate; they may grow a bit denser if growth factors are added to the medium. The cells lose their ability to move once the substrate is covered with a monolayer of cells.

   c. Cell line C grows quickly and is quite successful when the cells are suspended in soft agar. The cells are impervious to the addition of serum to their medium. Upon exposure to damaging radiation, most of the cells undergo apoptosis.

   d. Cell line D grows quickly and is impervious to DNA-damaging radiation, entering cycle after cycle of mitosis even after large doses of radiation. If transfected with DNA containing a gene for p53, these cells will undergo apoptosis.

Web Explorations

Vaccines against pathogenic invaders work by presenting the immune system with a harmless version of the pathogen that the body learns to recognize. Cancerous cells, however, are not foreign; they are normal cells that have undergone one or more mutations that cause them to grow out of control. Nonetheless, cancer cells synthesize proteins that differ slightly from their normal counterparts. Researchers have long believed that if they could teach the immune system to recognize these different proteins, they could stimulate the body's immune system to fight tumor cells and even develop a vaccine against cancer.

Using the World Wide Web, learn what progress has been made in the field of cancer vaccines. What types of vaccines are currently under development for cancer? What are DNA vaccines and how are they being used in the fight against cancer?

17

The Immune Response

Learning Objectives

When you have finished this chapter, you should be able to:

1. Understand the processes involved in both the innate immune response and the acquired immune response.
2. Describe the different cells that participate in the immune response, including the role of each type of cell in immunity.
3. Explain how antibodies are selected and amplified by the presence of an antigen.
4. Understand how the body can make millions of different antibodies from a limited number of different immunoglobulin genes.
5. Summarize the role of the major histocompatibility complex in immunity.
6. Explain how the immune system distinguishes "self" from "nonself."
7. Describe how different immune cells are activated, and how they respond to activation.
8. Give examples of autoimmune disorders and how they are being treated.

Key Terms and Phrases

immunity
innate immune response
natural killer cells
humoral immunity
immunoglobulin
T lymphocytes (T cells)
secondary immune response
tolerance
antigen presenting cell (APC)
helper T lymphocytes
epitope
somatic hypermutation
major histocompatibility complex (MHC)
graft-versus-host (GVH)

immune system
inflammation
interferon alpha
cell-mediated immunity
lymphocytes
antigen
memory cells
vaccination
cytokine
heavy chains
monoclonal antibody
histocompatibility antigens

costimulatory signal

immune response
complement
acquired immune response
antibodies
B lymphocytes (B cells)
clonal selection theory
autoantibodies
T-cell receptor (TCR)
cytotoxic T lymphocytes
light chains
signal sequences
MHC class I
MHC class II
JAK-STAT pathway

Reviewing the Chapter

I. Introduction
    A. The cells of the immune system engage in a type of "molecular screening" by which they recognize and attack foreign macromolecules.

    B. The weapons of the immune system:

        1. Cells that kill or ingest infected or altered cells.

     2.   Soluble proteins that can neutralize, immobilize, agglutinate, or kill pathogenic cells and viruses.

II. An Overview of the Immune Response

A. Innate immune responses occur immediately without previous contact with the microbe.
    1.   The response lacks specificity and concentrates defensive agents at the site of infection causing inflammation.
    2.   Natural killer cells induce apoptosis in cells infected with intracellular pathogens.
B. Acquired immune responses require a lag period to prepare for an attack against foreign agents.
    1.   This response is specific and occurs only in vertebrates.
    2.   There are two broad categories of acquired response:
        a.   Humoral immunity is carried out by antibodies which are secreted by B lymphocytes (B cells).
        b.   Cell-mediated immunity is carried out by T lymphocytes (T cells) that recognize and kill infected cells.

III. The Clonal Selection Theory As It Applies to B Cells

A. Antibodies are able to react with specific foreign substances, called antigens.
B. An antigen "selects" an antibody from preexisting antibodies capable of binding to it (clonal selection theory).
    1.   B cells become committed to produce one species of antibody.
    2.   B cells form antibodies in the absence of any antigens.
    3.   Thymus-independent antigens activate B cells, causing proliferation without the help of T cells.
    4.   Long-term immunity results from B lymphocytes that are retained as memory cells.
    5.   Immunological tolerance is developed in the body to prevent the production of autoantibodies.
C. Vaccination:
    1.   The first vaccine developed was for smallpox and produced immunity by stimulating T cells.
    2.   Most vaccines currently in use are B-cell vaccines.

IV. T Lymphocytes: Activation and Mechanism of Action

A. T cells, activated by clonal selection, interact with antigens through a surface protein called T-cell receptor.
    1.   T cells are activated by fragments of antigens that are displayed on the surface of antigen-presenting cells (APCs).
        a.   Dendritic cells ingest antigens by endocytosis.
        b.   Macrophages ingest antigens by phagocytosis.
        c.   These cells process and present the antigen to other cells.
    2.   T cells release cytokines that alter the activity of the target cell.
B. Two classes of T cells are distinguished by the proteins on their surfaces and their biological functions:
    1.   Cytotoxic T lymphocytes monitor cells in the body and cause abnormal cells to undergo apoptosis.
    2.   Helper T lymphocytes are regulatory cells that are activated by APCs.
    3.   Most B cells require the help of T cells to secrete antibodies.

V. Selected Topics on the Cellular and Molecular Basis of Immunity

A. The modular structure of antibodies:
    1.   Antibodies, or immunoglobulins, are globular proteins with heavy and light polypeptide chains.

2. There are five classes of immunoglobulins (IgA, IgD, IgE, IgG, IgM) defined by a unique heavy chain.
3. The assembly of antibodies from different combinations of light and heavy chains provides tremendous variety.
4. These chains contain regions of constant, variable, and hypervariable portions.
    a. Hypervariable regions allow for great diversity of antibody specificity.
    b. Constant regions vary between classes of immunoglobulins.
5. The epitope is part of the antigen that is complementary to the binding site on the antibody.
6. An antigen typically contains different epitopes which stimulate the production of a variety of different antibodies.
B. Antigen receptors on both B and T cells are encoded by genes produced by DNA rearrangement.
    1. Two separated genes (a C gene and a V gene) are combined through rearrangement to form one continuous gene that encodes one antibody chain.
        a. The process is catalyzed by V(D)J recombinase which joins V and J segments of the gene, and deleting the intervening DNA.
        b. Rearrangement is facilitated by signal sequences which are similar in V and J segments.
    2. Variability in polypeptide chains is achieved by:
        a. The variety of V and J exons in the DNA of the germ line.
        b. Varying the site at which J and V sequences are joined.
        c. The enzymatic insertions of nucleotides.
        d. Somatic hypermutation refers to a high mutation rate in V elements of B cells.
C. Antigen receptors are parts of large membrane protein complexes.
D. Antigens are presented to T cells following extensive intracellular processing.
    1. Histocompatibility antigens, including the major histocompatibility complex (MHC) determine compatibility between individuals.
    2. The MHC proteins hold fragments of antigen in place on antigen presenting cells (APC).
    3. The T cell receptor (TCR) interacts with an APC when it docks with MHC proteins.
    4. MHC class I proteins consist of one variable and one constant polypeptide chain.
        a. MHC I proteins display antigens originating within the cytosol of its cell.
        b. Proteases cleave the antigen into fragments which are transported into the ER and bind to MHC I molecules. This complex then becomes a membrane protein.
    5. MHC class II proteins consist of two variable chains and display fragments of exogenous antigens that are taken into a cell by endocytosis.
        a. MHC II molecules are joined to a Ii protein until they are directed into an endosome or lysosome.
        b. In the lysosome, the Ii is degraded and the MHC II binds antigen fragments that originated outside the cell.
    6. Cytotoxic T cells recognize their antigen in association with MHC I molecules.
    7. Helper T cells recognize their antigen in association with MHC II molecules.
E. How the body distinguishes self from non-self:
    1. T cell activation requires antigen fragments bound to self-MHC molecules.
    2. Nonself MHC molecules are treated as foreign and initiate a rejection response.

        a. A graft-versus-host reaction occurs in bone marrow transplants because the marrow-derived T cells are of non-host origin.

        b. Lymphoid stem cells from umbilical cord blood are less likely to react when used in place of bone marrow.

F. Lymphocytes are activated by cell surface signals:

   1. T cell activation requires interaction between its TCR and an MHC-peptide complex.

   2. Activation also requires a second signal called the costimulatory signal that is delivered by a second type of receptor.

   3. Activation of helper T cells by professional APCs:

        a. B7 proteins on APC cells interact with CD28 proteins on helper T cells in a costimulatory signal interaction.

        b. Only professional APCs are capable of costimulatory signaling; hence they are the only cells that initiate a helper T cell response.

        c. Helper T cell activation requires two signals, which protects other cells from autoimmune attack involving helper T cells.

   4. Activation of B cells and macrophages by T_H cells:

        a. Helper T cells bind to B cells whose receptors recognize the same antigen.

        b. B cells are activated by the interaction of cell surface proteins and by cytokines.

G. Signal transduction pathways utilized in lymphocyte activation:

   1. Several tyrosine kinases have been implicated in signal transduction during lymphocyte activation, including members of the Src family.

   2. Activation of these tyrosine kinases leads to a cascade of protein phosphorylation and activation of three distinct signal transduction pathways:

        a. activation of phospholipase C

        b. activation of Ras leading to the MAP kinase cascade

        c. activation of phosphatidylinositol 3-hydroxy kinase

   3. Lymphocytes produce cytokines that generate cytoplasmic signals that act on various intracellular targets.

        a. Cytokines use the JAK-STAT signal transduction pathway.

        b. The JAK-STAT pathway activates specific transcription factors.

        c. Interferon-stimulated response elements (ISRE) are DNA sequences that respond to JAK-STAT stimulated transcription factors and regulate the transcription of several genes

VI. The Human Perspective: Autoimmune Diseases

A. Autoimmune disorders result when the body mounts an immune response against part of itself.

   1. Some T and B cells possess receptors directed against the body's own proteins.

   2. Under normal conditions, self-reactive lymphocytes are prevented from becoming activated by a number of poorly understood mechanisms.

B. About 5% of the adult population suffers from an autoimmune disease including the following:

   1. Multiple sclerosis (MS) is a progressive neurological disease that results from an attack by both T cells and antibodies against components of the myelin sheath surrounding the axons of nerve cells.

   2. Insulin-dependent diabetes mellitus (IDDM) results from the autoimmune destruction of the insulin-secreting B cells of the pancreas.

   3. Grave's disease and thyroiditis attack different components of the thyroid but both lead to hyperthyroidism.

4. Rheumatoid arthritis is characterized by the progressive destruction of the body's joints due to a cascade of inflammatory responses.
5. Systemic lupus erythematosus (SLE) is not confined to a specific organ but often attacks tissues throughout the body, including the central nervous system, kidneys, and heart.

C. Many of these disorders indicate a strong genetic component to their development.
D. Studies have also demonstrated that environmental factors may also play a contributing factor.
E. New approaches for treatment of autoimmune diseases are being tested.
1. Immunosuppressive drugs
2. Induction of a return to a tolerant state so that the body stops producing autoantibodies and autoreactive T cells.
3. Interfering with the autoimmune response by administering antibodies.
4. Blocking the effect of inflammatory cytokines.
5. Blocking the body's ability to mount an immune response by immunization.

VII. **Experimental Pathways: The Role of the Major Histocompatibility Complex in Antigen Presentation**
A. Experiments in the 1970s indicated that cytotoxic T cells are MHC restricted.
1. Cytotoxic T cells kill only infected cells with the same H-2 genotype.
2. Research on the role of MHC in immunity was pursued as a result of these findings.
B. Experiments in the late 1970s and early 1980s indicated that T cells respond to antigen displayed on the surface of other cells.
1. Antigen had to be processed intracellularly before it was displayed.
2. Antigen processing was temperature sensitive.
3. Antigen processing was sensitive to metabolic poisons.
4. Antigen processing involved lysosomes.
C. Further studies implicated the MHC in T cell binding and antigen display.
1. Antigen-presenting cells could not bind T cells if treated with anti-MHC antibodies.
2. Processed antigen and MHC are isolated as a complex.
3. Radiolabeled processed antigen binds to the MHC isoform from the cells that process them.
D. The structure of both MHC I and the T cell receptor (TCR) have been characterized by X-ray crystallography.
1. Antigen peptides bind to a groove in the MHC I.
2. The TCR interacts with both the antigen peptide and MHC I.
3. The TCR recognizes both the foreign antigen peptide and the self MHC I.

Key Figure

Figure 17.9   Highly simplified, schematic drawing showing the role of T_H cells in antibody formation.

Questions for Thought:

1.   Does this drawing depict humoral immunity, cell-mediated immunity, both, or neither? Why?
2.   After this antigen has been conquered, what is the fate of the plasma cell?
3.   How would you add to this drawing to illustrate the cellular basis of long-term immunity?
4.   What changes would you make in this drawing to illustrate the effects of the HIV virus on the immune response? What cells would be affected? How do the symptoms of AIDS correlate with these steps in the immune response?

Review Problems

1. Give the significance (not the definition) of the following terms or phrases. Say what they do or why they are important. For example:
    antibodies: *recognize and bind to extracellular antigens and are, in turn, recognized by phagocytes or other effector cells of the immune system.*

    a. memory cells

    b. clonal selection

    c. professional APCs

    d. multiple myeloma as it pertains to research in immunology

    e. a single antigen may have several different epitopes

    f. somatic hypermutation

    g. Ii protein in the lumen of the rough endoplasmic reticulum

    h. all cells express MHC I proteins

    i. costimulatory signal

    j. JAK-STAT pathway

Chapter 17   255

2. Compare and contrast the following:

    a. innate vs. acquired immune response

    b. humoral vs. cell mediated immunity

    c. cytotoxic T cells vs. helper T cells

    d. antibodies vs. T cell receptors

    e. MHC I vs. MHC II

Multiple Choice

1. Which of the following is *not* characteristic of an innate immune response?
    a. inflammation
    b. increase in blood levels of specific antibodies
    c. increase in phagocytic cells at the site of infection
    d. activation of complement

2. List the order in which the events involved in humoral immunity occur:
    - I   B cell activation by T_H cell
    - II   B cell proliferation
    - III   Antigen processing and display by APC
    - IV T_H cell activation by professional APC
    - V   Antibody production and secretion

    a. III → IV → I → II → V
    b. III → II → IV → V → I
    c. V → III → II → I → IV
    d. IV → III → II → I → V

3. According to the widely-accepted clonal selection theory:
    a. all the antibodies you will ever be able to make are already present in lymphoid tissue.
    b. most of your lymphoid cells will never be recruited by antigen to produce antibodies.
    c. cells in the lymph tissues look alike under the microscope, but differ in their affinity for foreign proteins.
    d. all of the above.

4. Spleen cells from a mouse are separated and passed over column A containing beads coated with an antigen that the mouse has never encountered. The cells that are not retained by column A are poured over column B containing beads coated with an antigen the mouse *has* encountered in its life. What happens?
   a. A few cells stick to column A, but none stick to column B.
   b. Many cells stick to column A, but only a few stick to column B.
   c. No cells stick to column A, but many stick to column B.
   d. A few cells stick to column A, but many stick to column B.

5. You are studying a particular type of immune cell, but you are not sure which type it is. It is a relatively large, ovoid cell and has an extensive array of rough endoplasmic reticulum. Biochemical studies indicate that your cells lack any surface cluster designation (CD) proteins. What kind of cell might you have?
   a. plasma cell
   b. T_H lymphocyte
   c. natural killer cell
   d. dendritic cell

6. Of the following cell types, which ones function in both innate and acquired immune responses and as professional antigen presenting cells?
   a. macrophages
   b. T_c lymphocytes
   c. T_H lymphocytes
   d. B cells

7. T_H lymphocytes activate macrophages, Tc cells, and B cells, but what activates T_H cells?
   a. macrophages and/or dendritic cells
   b. T_H cells (i.e., self-activation)
   c. IL-1 and IL-2
   d. all of the above

8. The___?___portion of an antibody molecule determines the antigen binding specificity, whereas the___?___portion determines the class to which it belongs.
   a. heavy hypervariable, light constant
   b. IgM, IgG
   c. light and heavy variable and hypervariable, heavy constant
   d. light hypervariable, heavy variable

9. Antibodies present in breast milk are taken up by the infant's stomach via receptor-mediated endocytosis. Which portion of the antibody molecule is most likely recognized by the infant's receptors?
   a. C_H
   b. V_L
   c. V_H
   d. hypervariable

10. Which of the following is *not* a source of genetic diversity in immunoglobulin genes?
    a. multiple V exons and J exons in the DNA germ line
    b. variability in V-J joining
    c. genetic transposition
    d. enzymatic insertion of nucleotides by deoxynucleotide transferase

11. The genetic rearrangements that occur when antibody-producing cells mature are
    a. temporary; the original genetic sequences are restored after antibody production ceases.
    b. phenotypic; they do not involve the DNA.
    c. permanent; a cell that has undergone genetic rearrangement is forever limited in the kinds of antibodies it can produce.
    d. nonexistent; they are an artifact of the techniques used to study antibody production.

12. The significance of class switching is that the B cell lineage can
    a. recognize the same antigen but perform different effector functions.
    b. perform the same effector function on different antigens.
    c. convert to a T cell lineage.
    d. all of the above.

13. The B cell receptor is to the intact antigen as the T cell receptor is to ___?___.
    a. humoral immunity.
    b. cell-mediate immunity.
    c. processed antigen fragments.
    d. the intact antigen, also.

14. Professional antigen presenting cells have___?___, whereas most other cells have___?___.
    a. MHC I; MHC I and II
    b. MHC II; MHC I
    c. MHC I and II; MHC II
    d. MHC I; MHC II

15. What is the purpose of antigen-presenting MHC proteins on cells that are not part of the immune system?
    a. They evolved to alert the immune system to the presence of intracellular pathogens.
    b. They evolved to prevent tissue transplantation between individuals.
    c. They evolved to rid the cell of endogenous protein.
    d. They have no known function in non-immunological tissues.

16. The role of the Ii protein in class II MHC complexes is
    a. to display "self" proteins on the surface of APCs.
    b. to target the MHC complex to the Golgi.
    c. to provide a substrate for lysosomal enzymes.
    d. to prevent the MHC from binding an endogenous peptide before it encounters an antigen fragment.

17. When you meet someone you find attractive, you might speak of having a certain "chemistry" with that person. What real chemistry might underlie your instinctive attraction?
    a. You share certain genes in common with that person.
    b. Your different MHC genotypes would maximize the chances that any offspring you produce could recognize and fight pathogens.
    c. Your similar MHC genotypes would ensure that you could donate tissue to that person in the event of an emergency.
    d. You share the same abilities to make antibodies.

18. Which of the following is *not* a signal pathway between an activated B cell and a T_H cell?
    a. CD40 and CD40 ligand
    b. cytokines
    c. neurotransmitters
    d. neither a nor b are signal pathways.

19. Of the following intracellular cascades invoked by signal transduction, which is novel to activation by cytokines?
    a. ras and the MAP kinase cascade
    b. phosphatidylinositol 3-hydroxy kinase (PI(3)K) and membrane-bound lipid messengers
    c. phospholipase C and the IP3 and DAG pathway
    d. the JAK-STAT pathway for activating tyrosine kinases

20. Strep throat is a bacterial infection that can be followed several weeks later by rheumatic fever, an autoimmune disorder affecting the heart. How can the bacterial infection lead to the autoimmune disorder?
    a. Antibodies against strep cross react with a glycoprotein on heart valves.
    b. The heart valves are weakened by the massive immune response to the bacterium.
    c. Strep bacteria escape the immune system and take up residence in the heart.
    d. Strep stimulates the heart valves to make antibodies.

Problems and Essays

1. Vaccination and passive immunization are two ways of protecting against attack by pathogens, but they are not equally effective. In the figure below, label the line that illustrates the effects of vaccination on antibody levels and the line that illustrates the effects of passive immunization. Explain the differences between these two lines.

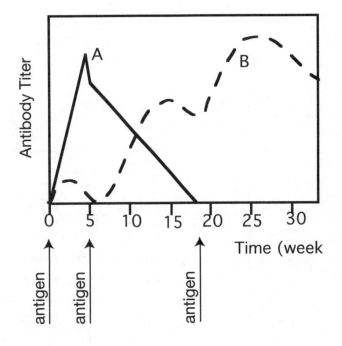

2. Of the experimental results described below (some real, some contrived to test your knowledge), mark with an S the ones that support the clonal selection theory, mark with an I the ones that are inconsistent, and mark with an N the ones that are not relevant to the clonal selection theory.

    a._____A population of lymphocytes from an adult mouse spleen is passed through a column of beads coated with antigen X, an antigen the mouse has not encountered before. None of the cells in the population that passes through without being retained by the column is able to make antibodies to antigen X.

    b._____Newborn mice exposed to antigen Y during the first few hours of life will exhibit tolerance toward antigen Y later in life.

c._____When a mouse is exposed to an antigen, the level of circulating B cells increases. The increase is shown to be due to proliferation of a single activated parent B cell.

d._____When a highly radioactive antigen Z is added to a population of lymphocytes, all cells that bind the antigen are killed by the radioactivity. The cells remaining in the population can be induced to make antibodies to a non-radioactive form of antigen Z.

e._____During the lifetime of an individual, only a very small percentage of mature B cells in the lymph tissue ever differentiate into plasma cells.

3. You have isolated a mixed population of lymphocytes from the spleen of an adult mouse, and now you wish to separate them according to cell type. To accomplish this, you do a procedure known as "panning for cells." You coat the bottom of a petri dish with antibodies to a specific cell surface protein. Next you pour the mixture of cells into the coated dish. After a few minutes, you pour off any cells that have not adhered. In the chart below, the first column lists the different specific antibodies you used to coat the petri dish. Indicate which cells will adhere under each of these different conditions.

Antibody	B cells	Helper T cells	CTL	Macrophages
anti-MHC II				
anti-CD4				
anti-CD 8				
anti-TCR				
anti-MHC I				
anti-IgG				
anti- CD28				
anti-B7				

4. Once a B cell is committed to producing specific antibodies, it may switch the class of immunoglobulin that it produces, for example from the IgM class to the IgG class. Design an experiment to determine if class switching involves genetic rearrangement.

5. You and your professor are doing an experiment using mice to measure primary and secondary immune responses to injections of antigen Z. Your professor collected the following values indicating the level of anti-Z antibodies in the blood of three different groups of mice, group A, group B, and group C.

Serum levels of anti-Z antibody.

Week	A	B	C
1	155	0	97
2	158	0	266
3	157	0	32
4	155	0	12
5	156	0	10,899
6	158	0	9,766

Unfortunately, the notebook in which he listed the experimental protocols for the different groups was destroyed when it fell into a sink full of soapy water. Your task is to reconstruct the experimental treatment given to each of the three groups of mice based on what you know about primary and secondary immune responses, and the data above.

Group 1 Treatment:_____
Explain:

Group 2 Treatment:_____
Explain:

Group 3 Treatment:_____
Explain:

6. Genetically identical mice from strain 1 were infected with virus A. After one week, cytotoxic T cells (CTL) were recovered from their spleens and tested for their ability to kill cultured fibroblast cells. For each of the following experimental conditions, predict whether the CTL will kill the fibroblasts (K), or not kill the fibroblasts (NK). Give a one or two sentence explanation for the result you predict.

a._____The fibroblasts were also taken from strain 1 and were infected with virus A.
Explain:

b._____The fibroblasts were also taken from strain 1, but were not infected with a virus.
Explain:

c._____The fibroblasts were taken from strain 2 and were infected with virus A.
Explain:

d._____The fibroblasts were taken from strain 1 and were infected with virus B.
Explain:

e._____The fibroblasts were taken from strain 1, but with a mutation in the gene at the MHC I locus. The fibroblasts were infected with virus A.
Explain:

Web Explorations

One of the goals of the massive international effort to solve the world-wide AIDS crisis has been to develop a vaccine for the HIV virus that causes AIDS. These research efforts, however, have been hindered by several characteristics of the HIV virus itself. One problem is the mutation rate of the viral genome; the virus' RNA genome mutates so rapidly that antibodies developed against one strain are impotent against a mutated strain. In one vaccine trial held at Harvard Medical School, several macaque monkeys given an attenuated form of the virus as a vaccine developed AIDS when the virus mutated into a virulent form.

Several other problems have plagued researchers searching for an HIV vaccine. Using the World Wide Web, learn about these problems, as well as the most recent advances in HIV vaccine research. What antigenic properties of the virus are most promising for vaccine development? In the event that an AIDS vaccine becomes available, what social and political issues will need to be addressed to make the vaccine widely available?

18

Techniques in Cell and Molecular Biology

Learning Objectives

When you have finished this chapter, you should be able to:

1. Describe the features of microscopy.
   a. Understand resolution, magnification, and visibility (contrast), as well as the technologies that enhance each.
   b. Know which types of microscopy are best suited for different experimental applications.
2. Understand how radioisotopes are used as tracers. Describe the methods of tracing radioactivity in cells, including autoradiography and liquid scintillation spectrometry. Know what types of insights each method provides.
3. Define cell culture; distinguish between different types of cell cultures. Understand the advantages of working with cells in culture, as well as some of the disadvantages.
4. Describe some techniques used for the separation and purification of proteins and nucleic acids.
   a. Understand the principles of centrifugation and know when to apply the different types of centrifugation.
   b. Understand the principles of chromatography and gel electrophoresis, and know when these techniques are used.
5. Know how x-ray diffraction provides insights into the three-dimensional structure of macromolecules.
6. Give an overview of the most commonly used recombinant DNA techniques.
   a. Describe DNA cloning; know what vectors are, and how they are used.
   b. Define DNA library, cDNA, site-directed mutagenesis, and PCR. Appreciate the importance of these techniques in the development of modern molecular biology.
   c. Know how transgenic animals are used as models.
7. Understand how antibodies are used for visualization of different materials; define hybridoma, monoclonal antibody, immunofluorescence.

Key Terms and Phrases

light microscopy
objective lens
empty magnification
bright-field microscope
polarization microscopy
shadow casting
freeze-fracture replication
confocal scanning microscopy
alpha particle

refractive index
ocular lens
fluorescence microscopy
phase-contrast microscopy
differential interference contrast
transmission electron microscope (TEM)
scanning electron microscope (SEM)
isotopes
beta particle

angle of incidence
resolution
numerical aperture
serial sections
negative staining
freeze-etching
critical-point drying
radioactive isotopes
gamma radiation

autoradiography
cell culture
protoplast
cell-free systems
specific activity
isoelectric point
cation exchanger
gel filtration chromatography
tracking dye
SDS-PAGE
x-ray crystallography
Svedberg (S) unit
DNA cloning
in situ hybridization
cDNA
transfection
animal models
antiserum
monoclonal antibody
indirect immunofluorescence
direct immunofluorescence

liquid scintillation spectrometry
mass culture
differential centrifugation
protein purification
selective precipitation
high-performance liquid chromatography (HPLC)
ion-exchange chromatography
affinity chromatography
polyacrylamide gel electrophoresis (PAGE)
isoelectric focusing
spectrophotometry
nucleic acid hybridization
vector DNA
yeast artificial chromosomes (YACs)
bacterial artificial chromosomes (BACs)
site-directed mutagenesis
knockout mice
polymerase chain reaction (PCR)
Sanger sequencing method
Gilbert and Maxam sequencing method
dideoxyribonucleoside triphosphate (ddNTP)

half-life
clonal culture
homogenizer
fractionation
chromatography

anion exchanger
electrophoresis

x-ray diffraction
ultracentrifugation
recombinant DNA
replica plating
DNA libraries
expression cloning
transgenic animals
embryonic stem cells
DNA sequencing
hybridoma

Reviewing the Chapter

I. Introduction
   A. Research in cell biology requires complex instrumentation and techniques.
   B. Understanding the technology aids in understanding the cell.

II. The Light Microscope
   A. The light microscope uses the refraction of light rays to magnify an object.
      1. The specimen is illuminated by rays from a light source.
      2. A condenser directs light toward the specimen.
      3. The objective lens collects light from the specimen to form an enlarged, real image.
      4. The ocular lens forms an enlarged, virtual image used by the eye.
   B. Resolution:
      1. Resolution is the ability to see two neighboring points as distinct entities.
      2. Resolution is limited by the wavelength of light.
      3. The numerical aperture is a measure of the light-gathering qualities of a lens.
      4. The resolution of a light microscope is 200 nm, sufficient for seeing large organelles.
      5. Lens aberrations affect resolving power.
   C. Visibility:
      1. Visibility requires that the specimen and the background have different refractive indices.
      2. Translucent specimens are stained with dyes.
   D. Phase-contrast microscopy:
      1. The phase-contrast microscope utilizes interference to convert the differences in the refractive index of some parts of a specimen into differences in light intensity.
      2. Differential interference contrast (DIC) optics give a three-dimensional quality.

E. Fluorescence microscopy (and related fluorescence-based techniques):
   1. Fluorochromes release visible light upon the absorption of UV rays.
   2. Fluorochrome stains cause cell components to glow.
       a. Fluorochrome-conjugated antibodies are used to locate specific cellular structures.
       b. The gene for green fluorescent protein (GFP) from jellyfish can be recombined with genes of interest in model organisms.
           (1) GFP is expressed with the host gene of interest.
           (2) GFP is used to follow the fate of the gene of interest.

   3. Fluorescence resonance energy transfer (FRET) uses fluorochromes to measure changes in distance between labeled cellular components.
F. Video microscopy is used to observe living cells; data can be processed by a computer.
G. Confocal scanning light microscopy:
   1. A laser beam is used to examine planes at different depths in a specimen.
   2. "Sections" can be viewed without cutting, as in CAT scans.
H. Preparation of specimens for light microscopy:
   1. A whole mount is an intact object, either alive or dead.
   2. Sections are very thin slices of an object.
       a. The tissue is fixated, embedded, and sliced.
       b. Procedures minimize alterations from the living state.

III. Transmission Electron Microscopy
   A. Transmission electron microscopes (TEMs) use electrons instead of light to form images.
       1. The limit of resolution of standard TEMs is 3 to 5 Å.
       2. The components of an EM:
           a. An electron beam from a tungsten filament is accelerated by high voltage.
           b. The electron beam is focused with a magnetic field.
           c. Electrons pass through the specimen and strike a fluorescent screen, creating an image.
           d. Differential scattering of electrons by the specimen creates the image.
               (1) Electron scattering (and visibility of an object) is proportional to the mass thickness of that part of the specimen.
               (2) Tissues are stained with heavy metals to obtain contrast.
   B. Specimen preparation for electron microscopy:
       1. Specimens must be fixed, embedded, and sectioned thinly.
           a. Glutaraldehyde and osmium tetroxide are common fixatives.
           b. Specimens are dehydrated prior to embedding.
           c. Epon is a common embedding material.
           d. Thin sections cut with glass or diamond knives are collected on grids.
       2. In negative staining, heavy metal diffuses into spaces between specimen molecules.
       3. Shadow casting coats a specimen with metal to produce a three-dimensional effect.
       4. Freeze-fracture replication and freeze-etching:
           a. In freeze-fracture replication, frozen tissue is fractured with a knife.
               (1) A heavy-metal layer is deposited on a fractured surface.
               (2) A cast of the surface is formed with carbon.
               (3) The metal-carbon replica is viewed in the TEM.
           b. In freeze-etching, a layer of ice is evaporated from the surface of the specimen prior to coating it with heavy metal.

IV. Scanning Electron Microscopy
    A. SEMs form images from electrons that have bounced off the surface of a specimen.
    B. Specimens for SEM are dehydrated by critical-point drying.
    C. Specimens are coated with a layer of carbon, then gold.
    D. The image in SEM is formed like an image on a television screen– it is indirect.
    E. SEM has a large range of magnification and depth of focus.

V. The Use of Radioisotopes
    A. Radioisotopes can be easily detected and quantified.
        1. A tracer is a substance that can be followed in cells.
        2. Properties of radioisotopes:
            a. Atoms that differ only in numbers of neutrons are isotopes of each other.
            b. Isotopes with an unstable combination of protons and neutrons are radioactive.
            c. Three forms of radiation can be emitted during atomic disintegration.
                (1) Alpha particles consist of two protons and two neutrons.
                (2) Beta particles are equivalent to electrons.
                (3) Gamma radiation is electromagnetic radiation or photons.
            d. The half-life of a radioisotope measures its instability; half of radioactive material disintegrates during its half-life.
        3. Liquid scintillation spectrometry:
            a. Scintillants absorb the energy of an emitted particle and release it in the form of light.
            b. Radiation of a tracer in a sample can be detected by measuring light emitted by a scintillant.
    B. Autoradiography:
        1. Autoradiography localizes radioisotopes within cells and tissues.
        2. A particle emitted from a radioactive atom activates a photographic emulsion.
        3. The location of the radioisotopes in the specimen is determined by the positions of the overlying silver grains in a photographic emulsion.

VI. Cell Culture
    A. Much of cell biology research is carried out on cultured cells.
        1. In a cell culture, cells can be obtained in large quantity.
        2. Most cultures contain only a single type of cell.
        3. Many different types of cells can be grown in culture.
        4. Cell differentiation can be studied in a cell culture.
    B. Cells in a culture require media that includes hormones and growth factors.
    C. Culture can be primary or secondary, depending on the origin of the cells.
        1. A primary culture consists of cells from the organism, usually an embryo.
        2. Cells in a secondary culture are derived from a previous culture.
    D. There are two types of primary culture– mass cultures and clonal cultures.
    E. Nonmalignant cells are capable of a limited number of cell divisions.
    F. Many types of plant cells can be grown in culture.

VII. The Fractionation of a Cell's Contents by Differential Centrifugation
    A. Differential centrifugation facilitates the isolation of particular organelles in bulk .
        1. Prior to centrifugation, cells are broken by mechanic disruption in a buffer solution.
        2. The homogenate is subjected to a series of sequential centrifugations.
        3. Organelles sediment into a pellet according to their size.
            a. Nuclei and remaining whole cells sediment first.

   b. Mitochondria, chloroplasts, lysosomes, and peroxisomes sediment next, followed by microsomal fragments..

   c. Ribosomes can be sedimented only in an ultracentrifuge.

 B. Cellular organelles isolated by centrifugation retain a high level of activity.

VIII. Isolation, Purification, and Fractionation of Proteins

 A. Protein purification involves the stepwise removal of contaminants.

  1. Protein purification can be either analytic or preparative.

  2. Purification is measured as an increase in specific activity of a protein.

 B. Selective precipitation:

  1. At low ionic strength, proteins tend to remain in solution (they are salted-in).

  2. At high ionic strength, protein solubility decreases.

  3. Ammonium sulfate is the most commonly used salt for protein precipitation.

 C. Liquid column chromatography:

  1. A mixture of dissolved components can be chromatographically fractionated as it moves through porous material.

   a. Components are fractionated between mobile and immobile phases.

   b. The molecule's rate of progress through the immobile phase is proportional to its affinity for the matrix.

   c. High-performance liquid chromatography (HPLC) has a high resolution.

  2. Ion-exchange chromatography:

   a. Proteins can be separated according to ionic charge.

   b. The isoelectric point is the pH value at which the number of positive and negative charges on the protein are equal.

   c. In ion-exchange chromatography, proteins with charged groups associate with the column matrix.

  3. Gel filtration chromatography:

   a. Gel filtration separates proteins by molecular weight.

   b. A gel filtration column is packed with cross-linked polysaccharides of different porosity.

   c. Proteins that are small enough to enter the pores are eluted from the column last.

  4. Affinity chromatography isolates one protein from a mixture using a specific ligand.

  5. Determining protein-protein interactions.

   a. Affinity chromatography can establish what proteins interacts with other proteins.

   b. Antibodies establish protein interactions by coprecipitation.

   c. Yeast two-hybrid system relies on recombinant DNA of a two-domain transcription factor:

    (1) A DNA binding domain is linked to the gene for one protein—the "bait" protein.

    (2) An activation domain is linked to genes encoding possible proteins that interact with the "bait."

    (3) A reporter gene (lacZ) is only expressed when the bait and its partner interact, bringing together the two domains of the transcription factor.

 D. Polyacrylamide gel electrophoresis:

  1. Electrophoresis is based on the fact that proteins can migrate in an electric field.

   a. In polyacrylamide gel electrophoresis (PAGE), proteins are driven by a current through a gel composed of cross-linked acrylamide.

   b. Movement of proteins depends on molecular size, shape, and charge density.

      c. The position of the proteins in the gel can be visualized by staining, autoradiography, or Western blot.

      d. SDS-PAGE:

          (1) PAGE is often done in the presence of the negatively charged detergent sodium dodecylsulfate (SDS).

          (2) In SDS-PAGE, proteins are separated on the basis of their molecular weight only.

2. Isoelectric focusing:

    a. In isoelectric focusing, the gel consists of a mixture of ampholytes that have varying ratios of positive and negative charges.

    b. Electrophoretic movement of ampholytes establishes a pH gradient.

    c. A protein stops when it reaches the pH equal to its isoelectric point.

3. Two-dimensional gel electrophoresis separates proteins on the basis of both isoelectric point and molecular weight.

4. Mass spectrometry identifies proteins

    a. Protein fragments are converted to ions and separated on the basis of mass and charge.

    b. Fragments are compared to large protein databases for identification.

IX. Determination of Protein Structure by X-ray Diffraction and Electron Crystallography

A. X-ray diffraction requires protein crystals.

B. Crystals are hit with x-rays; scattered radiation is collected by a photographic plate.

C. The diffraction pattern provides information about the structure of the protein.

D. X-ray diffraction is useful in the study of both proteins and nucleic acids.

X. Purification and Fractionation of Nucleic Acids

A. DNA purification procedures differ from protein purification procedures.

1. To obtain DNA, nuclei are isolated and lysed.

2. DNA is separated from contaminating materials (RNA and protein).

B. Separation of DNAs by gel electrophoresis:

1. PAGE is used for the separation of small DNA and RNA molecules. Large ones are separated in agarose gels.

2. Nucleic acids are separated on the basis of their respective molecular weights.

XI. Measurements of Protein and Nucleic Acid Concentrations by Spectrophotometry

A. The amount of protein or nucleic acid in a solution can be found by measuring the amount of light absorbed by the solution.

1. The light-absorbing capacity of a solution is read using a spectrophotometer.

2. Absorption of light is directly proportional to the number of absorbing molecules present in the light path.

B. Tyrosine and phenylalanine absorb light in the ultraviolet range.

XII. Ultracentrifugation

A. Sedimenting particles are affected by two opposing forces– diffusion and centrifugal force (which is dependent on particle density).

1. Large proteins diffuse slower than smaller ones.

2. Sizes of organelles and macromolecules can be expressed in Svedberg (S) units.

3. There are two types of ultracentrifuges.

    a. Preparative centrifuges are used to purify components for further use.

    b. Analytic ultracentrifuges allow one to follow the progress of the substances during the centrifugation.

B. Sedimentation behavior of nucleic acids:

1. Nucleic acid analysis involves the study of sedimentation patterns in an ultracentrifuge.

a. In velocity sedimentation, nucleic acid molecules are fractionated in a sucrose gradient on the basis of nucleotide length.

b. In equilibrium (isopycnic) sedimentation, nucleic acids are fractionated according to their buoyant density in a cesium gradient.

C. Nucleic acid hybridization:
1. Nucleic acid hybridization is based on the ability of two complementary DNA strands to form a double-stranded hybrid.
2. Nucleic acid hybridization methods:
   a. Two populations of single-stranded nucleic acids are incubated together.
   b. One population of single-stranded nucleic acids is usually immobilized.
   c. Nucleic acids that are complementary to the immobilized ones form double-stranded complexes.
   d. Nonhybridized single-stranded nucleic acids are washed away.
3. The Southern blot technique is based upon nucleic acid hybridization.
4. Nucleic acid hybridization can be used to determine the degree of similarity in nucleotide sequence between two samples of DNA.

XIII. Recombinant DNA Technology

A. Restriction endonucleases:
1. Restriction endonucleases cut DNA at specific sequences.
2. Restriction endonucleases are used to dissect genomes into precisely defined fragments for further analysis.
3. Restriction maps are complete diagrams of the fragments that result from digestion of a genome by specific restriction endonucleases.

B. Formation of recombinant DNAs:
1. DNA is first cut with restriction enzymes.
2. Recombinant DNAs can be formed in several ways, such as creating "sticky ends" with restriction enzymes.
3. The two components of a recombinant DNA are linked using DNA ligase.

B. DNA cloning:
1. DNA cloning produces large quantities of a specific DNA segment within host cells.
2. Bacterial plasmids and bacterial virus lambda (λ) are two commonly used vectors.
3. Cloning eukaryotic DNAs in bacterial plasmids:
   a. Plasmids used for DNA cloning are modified forms of the wild type.
      (1) Cloning plasmids contain a replication origin.
      (2) Cloning plasmids usually carry genes for antibiotic resistance.
   b. Recombinant plasmids are introduced into bacterial cells by transformation.
   c. Plasmid-containing bacteria are selected by treatment with antibiotics.
   d. Cells containing various plasmids are grown into separate colonies which can be screened for the presence of a particular DNA sequence.
      (1) Replica plating produces dishes with representatives of the same bacterial colonies in the same position on each dish.
      (2) In situ hybridization uses a labeled DNA probe to localize the colony possessing the desired DNA fragment.
4. Cloning eukaryotic DNAs in phage genomes:
   a. Recombinant DNA molecules are packaged into lambda phage heads.
      (1) Lambda phage can take inserts up to the size of 25 kb.
      (2) Sites of phage infection are identified as plaques in a bacterial "lawn."

(3) DNA fragments are identified by the same techniques as those used for cloning of recombinant plasmids.
   b. Formation of a DNA library:
      (1) A DNA library is a collection of DNA fragments representing the entire genome of an organism.
      (2) DNA fragments for the library can be obtained by cutting genomic DNA with restriction enzymes.
      (3) Cleaving of genomic DNA is random, which generates overlapping fragments.
      (4) Overlapping fragments are useful for chromosome walking.
5. Cloning larger DNA fragments in specialized cloning vectors:
   a. Yeast artificial chromosomes (YACs) can accommodate large DNA inserts.
   b. Bacterial artifical chromosomes (BACs) also accept large DNA inserts, and can be rapidly grown to large numbers.
6. cDNA libraries:
   a. The isolation of genomic fragments allows study of the genome. Topics include:
      (1) regulatory sequences flanking the coding portion of a gene.
      (2) noncoding intervening sequences.
      (3) the structure of multigene families.
      (4) the evolution of DNA sequences.
      (5) transposable genetic elements.
   b. Coding regions of a gene can be studied using cDNAs, synthesized by reverse transcriptase using mRNA as a template.
7. In expression cloning, cDNA is inserted next to a strong bacterial promoter, which allows foreign DNA to be transcribed and translated during the infection process.
C. Chemical synthesis and site-directed mutagenesis:
   1. Polynucleotides can be chemically synthesized.
   2. Site-directed mutagenesis inserts genetic elements containing modified nucleotide sequences into host cells.
D. Gene transfer into eukaryotic cells and mammalian embryos:
   1. DNA can be introduced into eukaryotic cells by the process of transfection.
   2. Transgenic animal and plants:
      a. Transgenic animals allow scientists to determine the effects of overexpression of a particular DNA sequence.
      b. Genetic engineering can produce animal models used to study human diseases.
      c. The main goal of genetic engineering in plants is to improve the efficiency of both photosynthesis and nitrogen fixation.
   3. Determining gene function by gene elimination:
      a. Knockout mice are obtained by transfecting embryonic stem cells, introducing them into an embryo, and implanting the embryo into a female mouse.
      b. Embryonic stem (ES) cells are used to make knockout mice.
E. RNA interference:
   1. Specific mRNAs can be degraded in vivo by treating with small double-stranded RNAs containing part of the sequence of the target mRNA.
   2. Cells treated with RNAi cannot make the protein encoded in the target mRNA.
   3. RNAi is a cheaper technique that knockout mice for learning about protein elimination.
F. Enzymatic amplification of DNA by PCR:

1. Polymerase chain reaction (PCR) cheaply, rapidly, and extensively amplifies a single region of DNA, from a very small amount of template.
2. PCR utilizes the DNA polymerase from bacteria living in hot springs.
3. PCR can be used in criminal investigations.
  G. DNA sequencing:
1. Techniques developed by Sanger, and Gilbert and Maxam are widely used for sequencing nucleic acids.
2. Sanger-Coulson (dideoxy) sequencing technique:
    a. Four samples of identical single-stranded DNA molecules are obtained.
    b. DNA of each sample is incubated with a primer, DNA polymerase, four dNTPs, and a low concentration of dideoxyribonucleoside triphosphate (ddNTP), different one in each sample.
    c. DNA fragments of different lengths are synthesized in each sample, with synthesis terminating where ddNTP has been (randomly) incorporated.
    d. DNA fragments are separated by gel electrophoresis and the DNA sequence is read from the gel bands.
    e. The amino acid sequence of the protein is deduced from the nucleotide sequence.

XIV. The Use of Antibodies
  A. Antibodies are highly specific proteins produced by lymphoid tissues in response to the presence of foreign materials.
  B. Preparation of antibodies:
1. A population of polyvalent antibodies can be obtained by repeated injections of a purified antigen into an animal; the blood of that animal serves as a source of the antiserum.
2. Univalent antibodies are produced by descendants of a single antibody-producing cell.
    a. Antibody-producing cells do not grow and divide in culture.
    b. Fusion of a normal antibody-producing lymphocyte and a malignant myeloma cell will create a viable hybridoma cell that can produce large amounts of a monoclonal antibody.
  C. Antibodies can be conjugated with a fluorescent substance that allows visualization of antigens.
1. In direct immunofluorescence, antibodies with bound fluorescent molecules bind to antigens. The binding sites are then visualized under a fluorescence microscope.
2. In indirect immunofluorescence, cells are first incubated with unlabeled antibodies, and then with labeled secondary antibodies which are directed against the primary antibody.

Key Figure

Figure 18.38.   An example of DNA cloning using bacterial plasmids.

Questions for Thought:

1.   To make a recombinant DNA molecule, it is necessary to cut both the donor DNA and the recipient DNA with the same restriction enzyme. Why?

2.   Why are the plasmids containng human DNA grown in bacteria and not human cells, such as those from which they came?

3.   What genetic elements must reside on the plasmid DNA in order for this technique to succeed? How are bacterial cells containing the plasmid selected?

4.   Are there ethical questions raised by recombinant DNA technology? What are they?

Review Problems

Short Answer

1. Give the significance (not the definition) of the following terms or phrases. Say what they do or why they are important. For example:
    light microscope: *enabled biologists to discover the existence of cells. This is still an important instrument for cell biologists.*

    a. microscopic resolution

    b. interference microscopy

    c. electron microscopy

    d. autoradiography

    e. cell culture

    f. two-dimensional gel electrophoresis

    g. nucleic acid hybridization

    h. yeast artificial chromosomes

    i. animal models for human disease

    j. polymerase chain reaction

2. Compare and contrast the following:

    a. magnification vs. resolution

    b. transmission electron microscopy vs. scanning electron microscopy

    c. differential centrifugation vs. density-gradient centrifugation

    d. cloning eukaryotic DNA in bacterial plasmids vs. in phage genomes

    e. cloning genomic DNA vs. cloning cDNA

Multiple Choice

1. A light microscope has an objective lens with a magnification of 40x and an ocular lens with a magnification of 10x. What is the total magnification of the image?
    a. 40x
    b. 50x
    c. 400x
    d. 450x

2. Which of the following does not contribute to the resolution attainable by a microscope?
    a. the wavelength of the light
    b. the magnification of the objective lens
    c. the refractive index of the material between the specimen and the lens
    d. the numerical aperture of the lens

3. Under which of the following conditions would the visibility of an object under a microscope be lowest?
    a. The object has the same refractive index as does the medium.
    b. The object and the background each bend light differently.
    c. The object diffracts some, but not all, of the light rays that hit it.
    d. The object absorbs some, but not all, of the light rays that hit it.

4. You are interested in reconstructing the three-dimensional shape of chromosomes in living cells that are in the process of mitosis. Which microscopic technique will you choose?
    a. freeze-fracture microscopy
    b. confocal scanning microscopy
    c. light microscopy
    d. scanning electron microscopy

5. You wish to find the intracellular compartments that contain high calcium concentrations. Which of the following microscopic techniques will you choose?
    a. fluorescence microscopy
    b. polarization microscopy
    c. phase-contrast microscopy
    d. transmission electron microscopy

6. Why are transmission electron micrographs never in color?
    a. Cellular structures have no color. They are all shades of gray.
    b. Color micrographs would be too expensive.
    c. Color film has not yet been invented.
    d. The photographic film detects electrons that are transmitted through the specimen rather than the different wavelengths of light that determine color.

7. You think you might have discovered a new organelle using electron microscopy, but it might be an artifact of your technique. What might you do to prove it is real?
    a. Look for the structure in many similarly prepared specimens from the same source.
    b. Look for the structure in many similarly prepared specimens from different sources.
    c. Look for the same structure in specimens prepared using different staining protocols.
    d. Look for the same structure in specimens prepared using different (or no) fixatives.

8. The half-life of tritium is 12 years. If you had some tritium and kept it for 24 years, what percentage of your original radioactive sample would be left?
    a. 12%
    b. 25%
    c. 75%
    d. none

9. You want to know if a culture of cells is in the process of DNA synthesis. You incubate your cells in the presence of radioactive thymidine to see if it is being incorporated into the DNA. What is the best technique to detect the labeled deoxynucleotide in nuclear DNA?
    a. autoradiography
    b. polyacrylamide gel electrophoresis
    c. agarose gel electrophoresis
    d. two-dimensional gel electrophoresis

10. The major difference between defined and undefined cell media used in cell culture is
    a. undefined medium is a recent development in cell culture technology.
    b. undefined medium is completely artificial.
    c. defined medium includes serum, lymph or other fluids from living sources.
    d. defined medium is free of serum, lymph or other fluids from living sources.

11. Which of the following would be the least promising source of cells for a primary cell culture?
    a. embryonic tissue
    b. actively growing malignant tumor tissue
    c. adult rat brain tissue
    d. plant protoplasts

12. You have homogenated and fractionated rat liver cells using differential centrifugation. Which fraction would have the fewest mitochondria?
    a. whole homogenate before any centrifugation steps
    b. the supernatant that results after the first low-speed centrifugation step
    c. the supernatant that results after a high-speed ultracentrifugation step
    d. None of the above fractions would have any mitochondria.

13. Which type of column chromatography separates proteins on the basis of molecular weight?
    a. ion-exchange chromatography
    b. gel filtration chromatography
    c. affinity chromatography
    d. isoelectric focusing

14. In gel electrophoresis, the tracking dye moves:
    a. more slowly than all the different molecules of the sample.
    b. at the same rate as all the different molecules of the sample.
    c. more quickly than all the different molecules of the sample.
    d. not at all.

15. Nucleic acid hybridization can be used as a measure of evolutionary relationships between species. Of the following statements about the DNA of related species, which is true?
    a. Closely related species form hybrid DNAs with relatively low melting temperatures.
    b. Closely related species form hybrid DNAs with relatively high melting temperatures.
    c. There is no correlation between DNA hybrid melting temperature and relatedness of species.
    d. One cannot create DNA hybrid molecules from DNAs of closely related species.

16. DNA fragments and DNA probes are to Southern blotting as___?___are to Northern blotting.
    a. DNA fragments and RNA probes
    b. RNA fragments and RNA probes
    c. RNA fragments and DNA probes
    d. protein fragments and DNA probes

17. The role of the vector DNA in DNA cloning is to:
    a. identify the source of DNA as foreign.
    b. identify the host cell that has taken up the one specific gene of interest.
    c. make the foreign DNA susceptible to digestion with endonucleases.
    d. carry the foreign DNA into the host cell.

18. Why does one need to make replica plates when screening for a specific DNA sequence among a large number of recombinant bacterial colonies?
    a. It may take several tries to positively identify the specific sequence of interest.
    b. The screening process requires several different steps, each of which must be done on a new colony of recombinants.
    c. One wants a living culture of recombinant cells available after screening, a process that destroys the cells.
    d. It is good science to replicate all experimental results.

19. What is a "gene gun"?
    a. a gun that fires DNA-coated pellets into plant cells
    b. a new technique using "explosively fast" enzymes to form recombinant DNA
    c. a weapon of mass destruction that exploits recombinant DNA technology for warfare
    d. none of the above

20. Why are heat-stable DNA polymerases from thermophilic bacteria required for the polymerase chain reaction?
    a. The heat-stable forms are the only ones that recognize all four deoxyribonucleotides.
    b. These enzymes amplify DNA in a reasonable amount of time.
    c. These enzymes are the most readily available forms of DNA polymerase in the world.
    d. These enzymes are stable enough to withstand the temperatures required to melt DNA.

Problems and Essays

1. You are studying subcellular structure with a microscope that has an angular aperture of 70° (θ = 70°) and you want to maximize your resolution.

   a. First calculate the resolution of your scope, using white light in air.

   b. Now place a blue filter between the light source and the specimen. The wavelength of blue light is 450 nm. Does this improve your resolution? If so, by how much?

   c. Add immersion oil (refractive index = 1.5), still using your blue filter. Now what is the resolution? Given your equipment, can you improve on this level of resolution?

Whole Homogenate

Centrifuge 600 x g

Supernatant 1

Pellet 1

Supernatant 1

Centrifuge 15,000 x g

Supernatant 2

Pellet 2

Supernatant 2

Centrifuge 100,000 x g

Supernatant 3

Pellet 3

2.

You have homogenized liver tissue and separated the subcellular fractions according to the centrifugation scheme shown to the left. You have saved small aliquots of each fraction.

   a. Using the aliquots of each fraction, you do enzyme assays specific for the enzyme succinic dehydrogenase (SDH). Recall that SDH is one of the enzymes of the TCA cycle. Assuming a perfect cell fractionation, which fractions will test positive for SDH activity?

   b. Which fraction will have the highest specific activity (activity per unit of protein) of SDH?

   c. Which fraction will have the most total protein?

   d. At the end of the fractionation, where would the ribosomes be found?

3.

time= 30 sec

time= 60 sec

time= 90 sec

time= 120 sec

One of the most elegant uses of radiolabeled tracers in cell biology occurred in the experiment done by Howard Dintzis in the early 1960s, where he determined the direction of proteins synthesis. Dintzis cultured reticulocytes actively engaged in synthesizing hemoglobin in the presence of radioactive amino acids. At short intervals, he collected samples of the cells and isolated the hemoglobin. The protein chains were digested with trypsin, a protease enzyme that cleaves polypeptides at the carboxyl sides of both arginine and lysine residues. For hemoglobin, this treatment yields a characteristic set of five fragments. labeled 1 through 5 in the figure. The amino acid sequence of hemoglobin was known, and hence the sequence of each trypsin fragment could be derived. After the fragments were separated, the radioactivity in each was measured. The diagram above summarizes Dintzis' experimental technique and his results. The hatched lines represent parts of the peptide that were radioactive at each of the time intervals. Based on these data, what did Dintzis conclude about the direction in which proteins are synthesized? Why?

4. Suppose you wish to pellet an 18S particle using ultracentrifugation. Assume the particle is at the top of a 7 cm centrifuge tube. The tube is centrifuged at 400,000 x g.

    a. What is the angular acceleration at 400,000 x g?

    b. At what velocity does the particle travel in this centrifugal field?

    c. How long will it take for the particle to pellet at the bottom of the centrifuge tube?

5. Traditional genetic approaches to problems of cell biology have followed the path of information from gene to protein. Biochemical approaches have traced proteins back to the DNA that encodes them, i.e., from protein to gene. This chapter has given you the basics to follow either path. Assume you are studying a problem relating to cell function. Briefly outline the basic steps you would take if you were starting with a gene and trying to deduce the function of the protein. Now outline the steps that you would take to get from a specific protein to a sequenced, isolated gene.

ddG   ddA   ddC   ddT

6. You wish to sequence a short nucleic acid (eight nucleotides) using the Sanger dideoxyribo-nucleoside method. After the fragments in each of your four test tubes have been separated and visualized with autoradiography, the patterns on the right are seen. What is the sequence of nucleotides in your fragment? Which end is the 5′ end and which is the 3′ end?

7. Name techniques you might use to answer each of the following questions:

   a. Does the DNA on the hair found at the scene of a crime match that of the defendant in a criminal trial?

   b. Where in a cell are the carbohydrates added to the glycoprotein hormones that are exported?

   c. Is it possible to find a cheap and plentiful source of human growth hormone for treating people with growth hormone deficiencies?

   d. Which amino acids in an enzyme play the most important roles in binding to the substrate?

Web Explorations

When most people think of the "biotech" industry, they think about drug development or genetic engineering for profit. While those industries do form an important part of biotech, many firms specialize in products that enhance or simplify basic cellular research, whether done in private sector or nonprofit laboratories. In this exercise, you will learn about some of the practicalities of using the many techniques you read about in Chapter 18.

Imagine you are a university research scientist engaged in basic research in some field of cell biology. Choose a research problem from among the many interesting areas covered in your textbook. On the basis of what you have learned in Chapter 18, what techniques will you apply to your problem and what kinds of information will these techniques provide? Now search the World Wide Web for commercial products that will enable you to carry out your experiments quickly and accurately. Find at least two companies on the web that provide products for your laboratory. Now use the company web sites to get an idea of the costs of doing different kinds of experiments.

Appendix:
Answers to Review Problems

Chapter One

Short Answer

1. Significance
   a. cells: *the smallest units to exhibit the fundamental property of life.*
   b. ATP: *stores chemical energy in cells in a readily available form.*
   c. model organisms: *provide a way for cell biologists to focus their studies on just a few organisms with the hope that understanding can be applied to issues of human importance.*
   d. prokaryotes and eukaryotes share the same genetic "language": *they share a common ancestry. In fact, eukaryotes evolved from prokaryotic ancestors.*
   e. nitrogen fixation: *makes otherwise useless nitrogen available for use by all cells by reducing N_2 gas to other forms, such as NH_3.*
   f. surface proteins on viruses: *bind to surface components of host cells. The viral surface proteins determine which cells will be infected by the virus.*
   g. genetic similarity between genes of viruses and their hosts: *provides evidence that viruses arose as bits of genetic material derived from the host genome that acquired some degree of autonomy from the host.*
   h. life cycle of Dictyostelium: *provides an excellent model system for studying the process of cellular differentiation in multicellular organisms.*
   i. bacterial conjugation: *provides a mechanism for prokaryotic cells to share DNA. Prokaryotes do not undergo meiosis, gamete formation, or fertilization, but are still able to recombine genetic material using this process.*

2. Compare and Contrast:
   a. prokaryote vs. eukaryote
      Together these two groups comprise the living world. They share a genetic language, some common metabolic pathways, and many structural features. Prokaryotes are the more primitive cells, and include the Archaebacteria and the Eubacteria. Eukaryotes comprise everything else, including the protists, fungi, plants, and animals. Prokaryotes are smaller. Prokaryotes have a single chromosome made entirely of DNA. They lack a true nucleus and membrane-delineated organelles, and they divide by simple fission. Eukaryotes have a true, membrane-bound nucleus containing chromosomes that are composed of both protein and DNA. They have membrane-bound organelles, endoplasmic reticulum, and cytoplasmic structural and contractile proteins. They divide by mitosis or meiosis, and undergo sexual reproduction. While both groups may have flagella, the flagella function quite differently in each.
   b. virus vs. virion
      "Virus" is a general term that refers to a group of non-cellular obligatory intracellular pathogens composed of either DNA or RNA, and a capsid, with or without a lipid membrane. "Virion" is a term describing a virus when it is outside the host.
   c. lytic infection vs. integrative infection
      These are two different mechanisms of viral infection. When a virus is lytic, the metabolism of the host cell is redirected to manufacture new viruses and assemble new virions. An integrative infection occurs when a virus integrates its DNA into the chromosome of the host, and either remains dormant or directs the cell to produce progeny virions without causing lysis of the host cell.
   d. Bacteria vs. Arachaea
      These are the two main groups, or domains, of prokaryotic cells. Both have living representatives, although the Archaea are believed to be more ancient than the bacteria. The closest living prokaryotic ancestor to the eukaryotes is believed to be an archaeon. The "true"

bacteria, in the domain Bacteria, are widespread and abundant in our everyday environment; the archaeons are generally found in extreme environments such as regions of high salt concentration or Earth's hot spots.

Multiple Choice

1.	a	2.	b	3.	c	4.	b
5.	a	6.	d	7.	d	8.	c
9.	d	10.	a	11.	d	12.	a
13.	d						

Problems and Essays

1. *These organelles cannot be maintained indefinitely when isolated from their cells of origin. They are incapable of reproduction outside of the cell, and rely upon the nucleus of the cell for many of the proteins that enable them to undergo respiration and photosynthesis. Therefore, they are not considered alive.*

2. *The process of differentiation, in which genetically identical cells develop into highly specialized cells, is poorly understood. The developmental pathway a cell takes depends upon signals it receives from the environment, and that, in turn, depends upon the position of that cell during the process of differentiation. External chemical signals can selectively activate some genes, while others remain dormant, thereby committing a cell to a particular developmental pathway.*

3. a. *Structures that increase surface area, such as infoldings or microvilli, and many mitochondria to fuel the active transport of substances across cell membranes.*
    b. *Structures that provide the equipment for protein synthesis, such as ribosomes and endoplasmic reticulum, and the vesicles that carry proteins from the site of synthesis to the membrane for release, such as Golgi and secretory vesicles.*
    c. *Many plastids, both for photosynthesis (chloroplasts) and for storing the products of photosynthesis.*
    d. *Prominent intracellular contractile proteins, and intracellular structure organized to direct the force of contraction in a specific direction.*

4. *Viruses are obligate intracellular parasites, and, as such, could not possibly have evolved before their hosts. Bacteria arrived on the scene about two billion years before the first eukaryotic cells, and over three billion years before humans. It is safe to say that the viruses that infect bacteria must have arisen long before those that infect humans. Further evidence that bacterial viruses are older than human viruses is found by comparing their structures. Bacterial viruses are typically more complex than human viruses, indicating that they have had more time to evolve.*

5. a. *The surface area of the cell is the area of each side (length x width) times the number of sides (= 6). SA = 100 μm x 100 μm x 6 = 60,000 μm^2. The volume of the cell is the length x width x height, or 100^3 = 1,000,000 μm^3. Thus, the SA/volume ratio =*

    $$SA/vol = 60,000 \text{ μm}^2 / 1,000,000 \text{ μm}^3 = 0.06 \text{ μm}^{-1}$$

    b. *No. One hundred twenty-five cubic cells would have a total surface area of 300,000 μm^2. The SA/volume would be 0.3. It would require dividing this cell into 125,000 individual cells, each 2 μm on a side to get a total surface area of 3,000,000 μm^2.*

6. *The common cold is caused by a virus. Unlike bacterial pathogens, viruses do not have their own machinery for replicating their genetic material. They rely entirely upon the host for replication. Thus drugs aimed at the bacterial replication process have no effect on viral replication. Drugs aimed at viral replication would likely have deleterious effects on the host organism.*

7. *The lipid envelope of the HIV virus is derived from the cell membrane of the host cell. The host cell produces new virions, which are released by budding and take some of the host membrane with them.*

8. *Photosynthesis in both cyanobacteria and green plants occurs by splitting water molecules, and releasing free oxygen. Photosynthesis in other prokaryotes occurs by an entirely different mechanism.*

9. a. *Possible pathogens include bacteria, viruses, or viroids. There is a remote chance that a prion could cause the corn disease, although no plant diseases have been attributed to prions so far.*
   b. *Bacteria can usually be cultivated from a homogenate of the infected tissue. Grind the infected roots in a sterile broth, and allow the broth to incubate for a day or two. Try infecting other corn plants with the bacteria that grow in the broth. If no bacterial pathogen is found, try filtering the root homogenate through a filter that is too small to allow bacteria to pass, but large enough to allow viruses to pass. A virus cannot be cultured in the absence of its host, so the filtrate must be applied to uninfected corn to determine its pathogenicity. Viroids are naked bits of RNA. A simple way to distinguish between a virus and a viroid would be to treat the filtered broth with an enzyme that attacks RNA and test it for pathogenicity. If the pathogen is a viroid, it will lose its ability to infect after such treatment.*

Chapter Two

Short Answer

1. Significance
   a. electronegativity: *describes how strongly a nucleus will attract electrons. In covalent bonds, the nucleus with the highest electronegativity attracts electrons more strongly, resulting in a polar bond.*
   b. free radicals: *implicated in both the cellular deterioration that accompanies aging, and the development of some diseases, including amyotrophic lateral sclerosis, cancer, and atherosclerosis.*
   c. carbon: *the atomic basis of life. Carbon's ability to form four covalent bonds with other carbons and/or atoms has resulted in the evolution of hundreds of thousands of organic molecules.*
   d. functional groups: *responsible for the physical properties and chemical reactivity of organic molecules.*
   e. ß(1 ⟶ 4) glycosidic linkages in cellulose: *a covalent bond that cannot be broken by the digestive enzymes of most animals. Cellulose is an abundant, energy rich molecule, but cannot be used as a source of energy by most animals.*
   f. amphipathic nature of phospholipids: *responsible for spontaneous bilayer formation making phospholipids ideal for their structural role as cell membranes.*
   g. protein primary structure: *ultimately determines all the properties of the protein, including its higher orders of structure, solubility characteristics, and function.*
   h. site-directed mutagenesis: *a technique that enables researchers to learn the roles of single amino acids in protein structure and function. This technique could ultimately be used to engineer or repair defective genes.*
   i. Anfinsen's experiments with ribonuclease: *the first series of experiments to show that the secondary and tertiary structure of the protein, ribonuclease, and its enzymatic activity, are based solely upon the primary structure.*
   j. molecular chaperones: *prevent hydrophobic interactions between both newly formed and heat-denatured proteins. If allowed to aggregate, proteins in these vulnerable states precipitate out of solution and become nonfunctional.*

2. Compare and Contrast:
   a. protein secondary (2°) structure vs. tertiary (3°) structure
   Both secondary and tertiary structure refer to the positions of atoms of a protein in space. Secondary structure of a protein refers to the spatial conformation of portions of a polypeptide chain and includes helical regions and sheet regions. Secondary structure is stabilized with weak bonds formed among the atoms of the peptide bond. Tertiary strucure describes the conformation of the entire protein and is stabilized by weak bonds among the side groups, or R groups, of the protein.

b. glycogen vs. starch
    *These are two polymers of glucose, both of which function as nutritional polysaccharides.
    Glycogen is stored in certain animal cells, e.g., the liver and muscle, and occurs in only one form:
    a highly branched, linear polymer. There are two forms of starch, the plant storage
    polysaccharide: amylose is an unbranched, helical molecule and amylopectin is a moderately
    branched linear molecule.*

c. saturated vs. unsaturated hydrocarbons
    *In saturated hydrocarbons, every carbon is covalently bound to the maximum number of
    hydrogen atoms; there are no double or triple carbon-to-carbon bonds. Unsaturated
    hydrocarbons are characterized by the presence of multiple bonds and thus are not fully reduced.
    Multiple bonds, unlike single bonds, restrict the free rotation of atoms. They also have higher
    bond energies.*

d. fat vs. oil
    *At the chemical level, these two lipids are essentially the same molecule; both are forms of
    triacylglycerols. In fact, an oil is actually a type of fat that is liquid at room temperature. The
    fatty acids that make up the triacylglycerols of oils have many cis double bonds. Double bonds
    introduce kinks into the linear structure of fatty acids and prevent them from packing together,
    thus lowering their melting point.*

e. purines vs. pyrimidines
    *These are the two classes of nitrogenous bases that occur in nucleotides. Purines are large, two-
    ringed structures that include the bases adenine and guanine. Pyrimidines are smaller, single-
    ringed structures that include the bases cytosine, uracil and thymine.*

Multiple Choice

1.	a	2.	c	3.	a	4.	b
5.	d	6.	d	7.	a	8.	b
9.	d	10.	d	11.	a	12.	c
13.	b	14.	a	15.	d	16.	b
17.	d	18.	a	19.	c	20.	a

Problems and Essays

1. *Solar radiation includes a broad range of wavelengths, including those at the low end of the
    electromagnetic spectrum. Most UV wavelengths have energy high enough to exceed the
    approximately 100 kcal required to break covalent bonds. Prolonged exposure to the full range of
    sunlight, including the UV component, causes tissue damage resulting from the breakage of covalent
    bonds. If the damage involves non-genetic biochemicals, the cells may die and be sloughed as in a
    bad sunburn. If cellular DNA is damaged but the cells do not die, errors in replication may occur
    when these cells divide, possibly leading to cancer. It is best and healthiest to avoid prolonged
    exposure to these high energy photons.*

2. *It takes one calorie of energy to raise one gram of water one degree Celsius, which, relative to other
    liquids, is actually quite high. Much of the energy that is absorbed is used to disrupt the
    intermolecular hydrogen bonds that form as a result of the polar, asymmetric nature of the molecule.
    Because energy is absorbed breaking weak bonds, the temperature of the water does not rise as much
    as that of another liquid might. In this sense, the temperature of a cell is buffered from changes in
    the temperature of the environment.*

3. a. *The description of three secondary structures and the fact that these proteins all bind calcium are
        characteristic of a motif.*
    b. *The long stretch of amino acids in these proteins, and their functional similarity, describe a
        domain. Domains are often attached to the rest of the protein by means of a "hinge" region.*
    c. *Again, the description of similar function with no mention of secondary structure would
        characterize the membrane-anchor region of these proteins as a domain.*
    d. *The zinc finger is a clearly defined set of secondary structures. This, and the common ability to
        bind DNA, would identify the zinc finger as a motif.*

4. (a) *E*; (b) *A and C*; (c) *A*; (d) *B*; (e) *B and F*; (f) *C, D, and E*; (g) *A is oleic acid, a fatty acid. B is α–D-glucose, a monosaccharide. C is a triacylglycerol, a fat. D is acetic acid, an acid. E is serine, an amino acid. F is adenosine monophosphate, a nucleotide.*

5.

This is just one example of a tripeptide composed of three hydrophobic amino acids. This one, H₃N–Leu–Ala–Val–COOH, would partition itself into the membrane, assuming the polar parts of the peptide were shielded from the lipid core. Any tripeptide with alanine, valine, leucine, isoleucine, tryptophan, phenylalanine, and methionine might partition itself into the membrane.

6. a. *Hydrogen bonds have bond energies in the range of 2 to 5 kcal/mole. Let's choose an average value of 3.5 kcal/mole. A mole of G–C pairs would have 3.5 x 3 = 10.5 kcal, and a mole of A–T pairs would have 3.5 x 2 = 7 kcal.*

    b. *The G–C rich strand would require more energy to denature, and therefore would be more stable.*

7. a. *Protein B has four prolines, each of which introduces a kink in the amino acid chain. B would not form an α helix.*

    b. *The hydrophobic residues in protein A, represented on the diagram below as circles, are scattered throughout the cylindrical α helix. Only protein C has all of the hydrophobic residues (represented as squares below) on one side of the α helix; protein C is the amphipathic protein.*

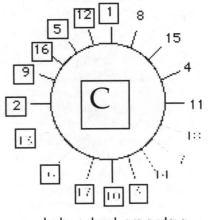

Chapter Three

<u>Short Answer</u>

1. Significance
    a. standard free energy difference, G°′: *provides a way of comparing the free energy changes of different reactions that do not naturally occur under the same conditions.*
    b. activation energy: *acts as a barrier preventing thermodynamically unstable reactants from reaching equilibrium. Makes possible kinetic stability, even in the face of thermodynamic instability.*
    c. kinetic stability: *enables substances to endure, although they may not be at their lowest energy level. Kinetic stability is made possible by the activation energy barrier.*
    d. structural homology between a substrate and an enzyme inhibitor: *characteristic of competitive inhibitors that bind to the active sites of enzymes.*
    e. transfer potential: *predicts which species will be the donor of an electron, atom or functional group (e.g.. phosphate) and which will be the acceptor.*
    f. enzyme specificity: *ensures that only the appropriate substrate will bind to an enzyme, and only an appropriate metabolic product will be made.*
    g. reciprocal plots of enzyme velocity vs. substrate concentration: *allows for more accurate estimates of important kinetic parameters, K_M and V_{max}, from velocity vs. substrate concentration data.*
    h. fermentation: *in anaerobic metabolism, rejuvenates the supply of NAD^+ for the continuation of glycolysis.*
    i. reducing power: *provides a measure of the cell's usable energy content.*
    j. feedback inhibition: *regulates a metabolic pathway early in the sequence of steps and according to levels of the ultimate product of the pathway.*

2. Compare and Contrast:
    a. equilibrium vs. steady state
        In both equilibrium and steady state conditions, the concentrations of reactants and products remain stable over time. The difference is that at equilibrium, stable concentrations represent identical levels of free energy (G = 0) and equal reaction rates in both the forward and reverse directions. In a steady state, there are differences in free energy levels (G 0), forward and reverse reaction rates are not equal, but concentrations remain the same because reactants are constantly being added and products are constantly being removed. No energy is required to maintain equilibrium, whereas a constant input of energy is required to maintain a steady state.
    b. kinetic stability vs. thermodynamic stability
        Substances that are kinetically and thermodynamically stable appear to remain unchanged over time. Thermodynamically stable substances are at equilibrium and, indeed, not changing. Kinetically stable substances are not at equilibrium and hence are changing, but the rate of change is so slow that it is imperceptible. The activation energy barrier enables thermodynamically unstable substances to remain kinetically stable.
    c. theoretically reversible reactions vs. essentially irreversible reactions
        All reactions are theoretically reversible; that is, they can proceed in either a forward or reverse direction according to the prevailing conditions under which they occur. Some reactions, however, are essentially irreversible; they only proceed in one direction because the conditions required to reverse them cannot or do not occur under natural conditions.
    d. NAD^+/NADH vs. $NADP^+$/NADPH
        These nicotinamide adenine dinucleotide coenzymes both function as electron "shuttles" in cells, alternately accepting electrons from or donating electrons to metabolic intermediates. The NAD^+/NADH redox pair functions primarily in catabolic pathways that yield ATP. The $NADP^+$/NADPH redox pair functions mainly to donate electrons to anabolic pathways.

e. competitive vs. noncompetitive inhibition

Both types of enzyme inhibition result in lowered rates of enzyme activity at subsaturating levels of substrate. Competitive inhibitors are substrate analogs that bind to the active site of the enzyme, thus competing with the real substrate and interfering with substrate binding (K_M). The effects of a competitive inhibitor can be overcome by adding high levels of substrate and raising the substrate/inhibitor ratio. Thus the V_{max} of an enzyme in the presence of a competitive inhibitor is unchanged. A noncompetitive inhibitor binds to a site other than the active site. It probably looks nothing like the substrate, and although it may have little effect on substrate binding (and hence K_M), the effect of the inhibitor cannot be overcome by adding saturating levels of substrate or raising the substrate/inhibitor ratio. Thus the V_{max} of the enzyme is lowered.

Multiple Choice

1.	b	2.	a. I	2.	b. I	2.	c. **D**
2.	d. **D**	2.	e. I	3.	c	4.	a
5.	d	6.	b	7.	a	8.	a
9.	d	10.	a	11.	d	12.	d
13.	b	14.	a	15.	b	16.	a
17.	c	18.	b	19.	d	20.	c

Problems and Essays

1. *For the reaction to proceed as written, $\Delta G'$ must be a negative number. Solve for the product/reactant ratio by setting $\Delta G' < 0$ as follows:*

$$\Delta G' < 0$$

$$\Delta G^{o\prime} + 2.303\, RT \log \frac{[glyceraldehyde\,3-phosphate]}{[dihydroxyacetone\,phosphate]} < 0$$

$$1800\ cal/\ mole + (2.303)(1.987\ cal/\ mol \bullet^{\circ} K)(298^{\circ} K) \log \frac{[glyceraldehyde\,3-phosphate]}{[dihydroxyacetone\,phosphate]} < 0$$

$$1364 \log \frac{[glyceraldehyde\,3-phosphate]}{[dihydroxyacetone\,phosphate]} < -1800\ cal/\ mol$$

$$\log \frac{[glyceraldehyde\,3-phosphate]}{[dihydroxyacetone\,phosphate]} < -\frac{1800}{1364}\ cal/\ mol$$

$$\log \frac{[glyceraldehyde\,3-phosphate]}{[dihydroxyacetone\,phosphate]} < -1.32\ cal/\ mol$$

$$\frac{[glyceraldehyde\,3-phosphate]}{[dihydroxyacetone\,phosphate]} < 0.048\ cal/\ mol$$

2. Solve for $\Delta G^{\circ\prime}$ as follows:

$$\Delta G^{\circ\prime} = -(2.303)\, RT \log K'_{eq} \qquad \text{where } K'_{eq} = 0.2818$$

$$= -(2.303)(1.987\ cal/\ mol\cdot{}^\circ K)(298^\circ K)\log 0.2818$$

$$= -(1364\ cal/\ mol)\log 0.2818$$

$$= -(1364\ cal/\ mol)(-0.5501)$$

$$= 750\ cal/\ mol \qquad or \qquad 0.750\ kcal/\ mol$$

3. Solving this problem requires the use of simultaneous equations. First, use the relationship between equilibrium and standard free energy to find the ratio of products to reactants at equilibrium. Next, you know that for every molecule of glucose 6-phosphate that is produced, one molecule of fructose 6-phosphate must be consumed. Therefore, the total concentration of both, at any stage of the reaction (equilibrium included), must be [glucose 6-phosphate] + [fructose 6-phosphate] = 0.50 M. Solve the problem using this information:

$$\Delta G^{\circ\prime} = -400\ cal/\ mol$$

$$-400\ cal/\ mol = -2.303\ RT \log K'_{eq}$$

$$-400 = -(2.303)(1.987\ cal/\ mol\cdot{}^\circ K)(298^\circ K)\log K'_{eq}$$

$$\frac{-400}{-1364} = \log K'_{eq} \qquad 0.2933 = \log K'_{eq} \qquad K'_{eq} = 1.96$$

$$also \quad K'_{eq} = \frac{[glucose\ 6-phosphate]}{[fructose\ 6-phosphate]} \qquad at\ equilibrium$$

$$and \quad [glucose\ 6-phosphate] + [fructose\ 6-phosphate] = 0.50\ M$$

$$so \quad [glucose\ 6-phosphate] = 0.50\ M - [fructose\ 6-phosphate]$$

$$thus \quad 1.964 = \frac{0.50 - [fructose\ 6-phosphate]}{[fructose\ 6-phosphate]}$$

$$1.96\,[fructose\ 6-phosphate] = 0.50\ M - [fructose\ 6-phosphate]$$

$$2.96\,[fructose\ 6-phosphate] = 0.50\ M$$

$$[fructose\ 6-phosphate] = \frac{0.50\ M}{2.96}$$

$$[fructose\ 6-phosphate] = 0.17\ M$$

$$[glucose\ 6-phosphate] = 0.50 - 0.17$$

$$[glucose\ 6-phosphate] = 0.33\ M$$

4. a. Using the same reasoning as in number (3) above, solve for [fumarate] as follows:

$$\Delta G^\circ = 0 = -2.303\ RT \log K_{eq}$$

$$0 = \log K'_{eq}$$

$$K'_{eq} = 1.0$$

$$thus \quad \frac{[fumarate][FADH_2]}{[succinate][FAD]} = 1.0\ at\ equilibrium$$

$$therefore\ [fumarate] = 0.005\ M$$

b. It will not change the answer at all. Enzymes have no effect on the equilibrium concentrations of products and reactants.

5. $\dfrac{-686.0 \ \text{kcal/mole glucose}}{7.3 \ \text{kcal/mole ATP}} \times 0.4 = 37.6$ mole ATP/mole glucose

6. $\dfrac{-2385 \ \text{kcal/mole palmitic acid}}{7.3 \ \text{kcal/mole ATP}} \times 0.4 = 131$ mole ATP/mole glucose

7. Fat has over three times more calories per mole than sugar. To gain weight, it is better to eat fats. To lose weight, it is best to stay away from fats.

8. a. *B*   b. *C*   c. *N*   d. *C*   e. *N*   f. *C*

9. a. *Lineweaver-Burk Plot of Kinetic Data:*

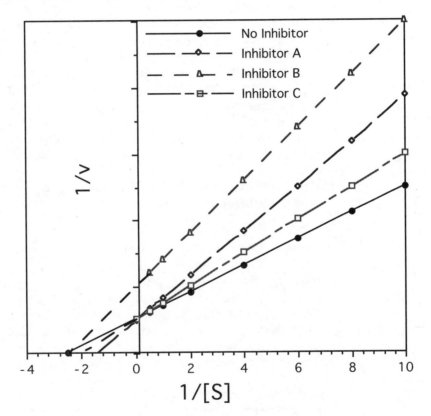

b. and c.

Kinetic Parameters	No Inhibitor	Inhibitor A	Inhibitor B	Inhibitor C
K_M	0.40	0.68	0.40	0.50
V_{max}	2.00	2.00	1.00	2.00
type of inhibitor	none	competitive	noncompetitive	competitive
why?		V_{max} unchanged	K_M unchanged	V_{max} unchanged

Chapter Four

<u>Short Answer</u>

1. Significance
    a. membrane leaflet asymmetry: *results in different functional properties of the inside and outside surfaces of the membrane. For example, the outside surface can recognize signal molecules via specific receptors. The inside surface may have enzymes, or proteins involved in anchoring cytoskeletal elements.*
    b. partition coefficient: *a measure of the solubility of a substance in both polar and nonpolar solvents. Substances with high partition coefficients readily enter and leave cells by simple diffusion.*
    c. flippase enzymes: *provide a mechanism for the exchange of phospholipids between the inner and outer leaflet of the cell membrane.*
    d. freeze-fracture and freeze-etching: *techniques that show the arrangement of membrane lipids and proteins, including microdomains.*
    e. detergents: *amphipathic substances that permit otherwise insoluble proteins, such as integral membrane proteins, to be solubilized in aqueous solutions, without denaturing.*
    f. hydropathy plots: *by assigning a hydrophobicity value to amino acids in a protein, these plots identify stretches of amino acids that are energetically stable within the lipid bilayer and therefore represent transmembrane domains..*
    g. chemical-gated channels: *respond to signal molecules such as hormones and neurotransmitters by opening ion channels and permitting ions to enter or leave cells.*
    h. Na^+-K^+, ATPase: *establishes and maintains the gradients of both the Na^+ and K^+ ions between the inside and outside of cells.*
    i. membrane potentials: *result from the unequal distribution of ions between the inside and outside of cells. Some cells, such as nerve, muscle and some endocrine cells, use membrane potentials as a source of energy.*
    j. myelin: *serves as an insulator of nerve cell axons. Myelin restricts action potentials to nodes along the length of the axon, resulting in rapid, saltatory impulse conduction.*

2. Compare and Contrast:
    a. the membrane surface facing inside the cell vs. that facing outside
    The inner and outer leaflets of the membrane differ in both lipid and protein composition, and consequently in function. The outer leaflet is richer in phosphatidylcholine, whereas the inner leaflet contains most of the phosphatidylethanolamine, phosphatidylinositol, and phosphatidylserine. The outer leaflet houses all of the glycolipids and glycoproteins, whereas no carbohydrate groups extend into the interior of the cell from the membrane. Protein asymmetry gives the membrane its functional attributes, including signal recognition (outer leaflet) and response (inner leaflet), cell recognition (outer leaflet), and membrane participation in cellular metabolism (inner leaflet).
    b. membrane-lipid mobility vs. membrane-protein mobility
    The fluid-mosaic model of membrane structure predicts that both lipids and proteins will move freely in the plane of the membrane. Actual studies of mobility, however, have shown that lipids tend to move more freely than proteins. The movement of many proteins is restricted by connections with intracellular structures, such as cytoskeletal components, whereas the movement of most lipids has no such restrictions.
    c. diffusion vs. facilitated diffusion
    Both these terms are used to describe the movement of substances across cell membranes down a concentration gradient– a process that does not require an input of energy. Diffusion implies movement through the bilayer without the help of a carrier, whereas facilitated diffusion requires a carrier molecule (usually a protein) that aids the substance in crossing the membrane barrier.
    d. membrane potential vs. resting potential
    These terms describe the voltage that can be measured between the inside and outside of a cell. Membrane potential is the voltage that occurs across the membrane of a nonexcitable cell. The

term resting potential applies to excitable cells, and implies that the cell can undergo changes in that potential related to the work of the cell; e.g., the action potential of nerve and muscle cells.

e. myelinated nerve cells vs. large-diameter nerve cells
Myelination and large diameters in nerve cells are different evolutionary strategies that increase the rate at which nerve cells carry impulses. Invertebrates have adopted the large-diameter strategy, but there are limits to the degree to which increasing diameter can speed up impulses. Vertebrates have evolved a myelin sheath that surrounds nerve cells– an adaptation that increases impulse conduction velocity by as much as twenty-fold.

Multiple Choice

1.	c	2.	b	3.	d	4.	b
5.	a	6.	b	7.	a	8.	a
9.	c	10.	b	11.	a	12.	b
13.	a	14.	d	15.	a	16.	a
17.	d	18.	a	19.	b	20.	a

Problems and Essays

1. (a) The regulation of metabolic pathways described in Chapter 2 relies partly on the maintenance of intracellular areas of high substrate and cofactor concentrations. Compartmentalization makes that possible by preventing the substrates and cofactors of specific metabolic pathways from being diluted by mixing with the entire cell's cytoplasm. (b) Resting potentials that are used by excitable cells to effect action potentials depend upon regions of highly regulated ion concentrations. Cellular compartmentalization makes that possible. (c) Throughout this book you will learn about cellular processes made possible by intracellular compartmentalization. Aerobic metabolism, described in Chapter 7, and photosynthesis characteristic of green plants, described in Chapter 8, are two such processes.

2. Identify a tissue that would serve as a good source of the enzyme of interest, e.g., the plasma membranes of nerve tissue. Solubilize the Na^+-K^+ ATPase from its native membrane using detergents. Reconstitute the enzyme into liposomes of simple and well-defined composition, then assay the enzyme for activity. By synthesizing a variety of liposomes with different amounts of saturated and unsaturated fatty acids, the fluidity of the liposome can be altered, and the enzyme can be tested under conditions of different fluidity. (To measure the fluidity of the liposomes directly, without having to infer it from the fatty acid composition, either FRAP or SPT techniques can be applied.)

3. a. C   b. C   c. N   d. C   e. N

4. a. P   b. I   c. LA   d. I   e. I   f. I

5. The equation that describes the energy required to move a substance into a cell is:

$$\Delta G = RT \ln \frac{[C_i]}{[C_o]}$$

but in this case we are interested in moving a substance out of a cell, against a concentration gradient. Therefore we must invert the concentration ratio:

$$\Delta G = RT \ln \frac{[C_o]}{[C_i]}$$

Now substitute for the concentration terms:

$$\Delta G = 2.303 \, RT \log \frac{3.00}{0.04}$$

$$\Delta G = 2.62 \, kcal/mole$$

6. The equilibrium distribution of molecules across a membrane results from the forces of both concentration gradients and electrical gradients (measured as a membrane potential). For

uncharged molecules, electrical gradients have no effect. For ions, however, membrane potentials may act in either the same direction as concentration gradients, or the opposite direction. If an ion is distributed across a cell membrane such that the concentration gradient and the electrical gradient are equal in magnitude but opposite in direction, then the equilibrium distribution of the ion may be represented by different concentrations across the membrane.

7. *First calculate the resting portential of the cell under three different conditions, where each of the three ions is assumed to be at equilibrium:*

ion:	Ψ^+	Ξ^+	Θ^-
inside cells	180	6	34
outside cells	21	244	51

$$E_\Psi = 2.303\frac{RT}{zF}\log_{10}\frac{[\Psi]_o}{[\Psi]_i} \qquad E_\Xi = 2.303\frac{RT}{zF}\log_{10}\frac{[\Xi]_o}{[\Xi]_i} \qquad E_\Theta = 2.303\frac{RT}{zF}\log_{10}\frac{[\Theta]_o}{[\Theta]_i}$$

$$E_\Psi = 2.303\frac{RT}{zF}\log_{10}\frac{21}{180} \qquad E_\Xi = 2.303\frac{RT}{zF}\log_{10}\frac{244}{6} \qquad E_\Theta = 2.303\frac{RT}{zF}\log_{10}\frac{34}{51}$$

$$E_\Psi = -55 \text{ mV} \qquad\qquad E_\Xi = 95 \text{ mV} \qquad\qquad E_\Theta = -10 \text{ mV}$$

Now compare these calculated values with the actual resting potential of +100 mV. The ion Ξ^+ is the one that is closest to its equilibrium potential, and hence probably the one to which the membrane is most permeable. Thus the answers to both (a) and (b) are Ξ^+. (c) The ion Ψ^+ is furthest from equilibrium and, if it were allowed to cross the membrane, would result in the greatest change in membrane potential. Thus the Loligotian action potential probably results from opening voltage-gated Ψ^+ channels.

Chapter 5
Short Answer
1. Significance
   a. evolutionary success of the cyanobacteria: *responsible for the accumulation of oxygen in the atmosphere, and the eventual success of organisms that could utilize oxygen as a final electron acceptor in metabolism.*
   b. cristae: *increase the inner mitochondrial membrane surface area, and thus the number of oxidative phosphorylation sites.*
   c. porins: *create pathways for all but the largest cytoplasmic components to enter the outer mitochondrial membrane.*
   d. spatial arrangement of carriers in the electron-transport chain: *expedites the transfer of electrons between components of the chain by minimizing the distance they must travel within the inner mitochondrial membrane.*
   e. only 20% of Δp is ΔpH: *permits the accumulation of a large electrochemical gradient without lowering the cytoplasmic pH to a level of acidity that would be detrimental to cytoplasmic proteins.*
   f. inner mitochondrial membrane is impermeable to H^+: *prevents the proton gradient established by electron transport from dissipating by diffusion back into the mitochondrial matrix.*
   g. levels of ADP in the mitochondria: *regulate the rate of oxidative metabolism.*
   h. fermentation: *regenerates NAD^+ in the absence of oxygen so that glycolysis can continue.*
   i. catalase in the peroxisomes: *rapidly breaks down potentially toxic H_2O_2 that results from many of the oxidase enzymes residing in these organelles.*

j. Na$^+$,K$^+$ ATPase from red blood cells can be made to synthesize ATP: *This observation, made under highly contrived conditions, provided evidence that an enzyme that catalyzes a reaction with a highly negative standard free-energy change can be made to catalyze the reverse reaction under the right cellular conditions. More specifically, an ATPase can be made to synthesize ATP from ADP and unbound P$_i$.*

2. Compare and Contrast:
   a. inner vs. outer mitochondrial membranes
   Both of these membranes surround the mitochondrion, but they differ in nearly every aspect of their structure, function, and origin. The outer membrane is highly permeable to even substances of large molecular weight (due to the presence of porins); the inner membrane is highly impermeable to most cellular molecules. The lipid composition of the outer membrane resembles that of the plasma membrane, whereas the inner membrane lipids are more characteristic of bacterial plasma membranes from which they are believed to have evolved. The outer membrane is involved in fatty acid, tryptophan, and epinephrine metabolism (among other things), and in recognizing proteins to be imported into the mitochondria. The inner membrane is involved in electron transport and the synthesis of ATP.
   b. "high energy" vs. "low energy" electrons
   All electrons are affiliated with some atom or molecule. A "high energy" electron is affiliated with a species that is a good reducing agent– one that readily transfers the electron to another compound or atom with the concurrent release of free energy. A low energy electron is affiliated with a compound or atom that has a low electron-transfer potential, and hence cannot readily be tapped as a source of free energy by a redox reaction with another species.
   c. substrate-level vs. oxidative phosphorylation
   Both of these terms describe the synthesis of ATP from ADP and a phosphate group. In substrate-level phosphorylation, the phosphate group is covalently bound to a donor, and that bond is broken when it is transferred to ADP. In oxidative phosphorylation, there is no phosphate donor. An ATP synthase catalyzes the synthesis of ATP from ADP and free inorganic phosphate using energy derived from a proton gradient . The term "oxidative" refers to the source of energy for making the proton gradient, i.e., the reduction of oxygen by organic reductants in the electron-transport chain.
   d. chemical vs. electrical gradient
   A chemical gradient is based upon a concentration difference between compartments, and the tendency for diffusion to lessen the difference. An electrical gradient occurs where there is a separation of charge between compartments. Like charges repel each other. Protons distributed across the inner mitochondrial membrane are subject to both types of gradients
   e. redox loop vs. proton pump
   These are two possible mechanisms that have been proposed for the translocation of protons across the inner mitochondrial membrane during electron transport. A redox loop occurs when an electron carrier picks up both protons and electrons, passes the electron on to a subsequent carrier, and dumps the proton into the intermembrane space. Redox loops require an alternation of proton-transporting and non-proton-transporting electron carriers. A proton pump moves protons across the membrane using the energy of electron transfer, and may undergo conformational changes to accomplish the task.

Multiple Choice

1. d	2. b	3. c	4. a
5. d	6. a	7. b	8. a
9. c	10. a	11. a	12. a
13. c	14. d	15. c	16. b
17. d	18. b	19. a	20. b

Problems and Essays

1. *(a) Size. The size of the typical mitochondrion is comparable to that of a typical bacterium.*

(b) Membrane-lipid composition. The inner mitochondrial membrane is rich in cardiolipin and devoid of cholesterol– features of many bacterial plasma membranes.
(c) The genome, and the transcriptional and translational machinery.
These features suggest that mitochondria were once free-living. In addition, the circular DNA and the ribosomes resemble prokaryotic features more so than eukaryotic ones.

2.   a.   acetyl CoA + $2H_2O$ + $3NAD^+$ + FAD + ADP + P_i $\Leftrightarrow$

$$2CO_2 + 3NADH + 3H^+ + FADH_2 + CoASH + ATP$$

b.   *Because the TCA cycle is, indeed, a cycle, for each acetyl CoA that consumes an oxaloacetate, there is one oxaloacetate that is produced from malate.*

3.   a.   ubiquinone$_{red}$ + 2 cytochrome c_{ox} $\Leftrightarrow$ ubiquinone$_{ox}$ + 2 cytochrome c_{red}

b. $\Delta G^{\circ\prime}$ = $-nF \Delta E'_o$
$$= -(2)\ (23.063\ kcal/V \bullet mol^{-1})\ (0.254\ V - 0.100\ V)$$
$$= -7.103\ kcal/mol$$

c.   *No. The 7.1 kilocalories released in the redox reaction are just shy of the 7.3 kilocalories required to synthesize one mole of ATP under standard conditions.*

4.   a.   *Under standard conditions, ubiquinone would act as the oxidant and cytochrome b would act as the reductant, donating electrons to ubiquinone. Ubiquinone has a higher redox potential than cytochrome b.*

b.   *Yes, the roles of oxidant and reductant could be reversed by changing the relative concentrations of the oxidized and reduced forms of one or both species. Increasing the concentrations of reduced ubiquinone and/or oxidized cytochrome b is one strategy. Decreasing the concentrations of oxidized ubiquinone or reduced cytochrome b is another.*

5.   a. $NADH + H^+ + 1/2 O_2$ $\Leftrightarrow$ $NAD^+ + H_2O$
$\Delta G^{\circ\prime}$ = $-nF \Delta E'_o$
$$= -(2)\ (23.063)\ (0.82 - (-0.32))$$
$$= 52.6\ kcal/mol$$

b.   *On the cellular scale, 52.6 kcal for each mole of NADH is a very large package of energy, about equal to the energy in an entire Oreo cookie. In a single mitochondrion, releasing that amount of energy would be akin to exploding a small stick of dynamite. By releasing the energy in smaller packets, using the carriers of the electron-transport chain, the cell can capture the free energy without suffering damage.*

6.

$$\Delta p = \Psi - 2.3\frac{RT}{F}\Delta pH$$

$$230\ mV = 0.8(230\ \ mV) - 2.3\frac{RT}{F}\Delta pH$$

$$46\ \ mV = -59\Delta pH$$

$$-0.78 = \Delta pH$$

where $\Psi = 0.8\Delta p$

and $\Delta p = 230\ mV$

and $2.3\frac{RT}{F} = 59\ mmV$

7.   a.   When inside out submitochondrial particles are engaged in electron transport, the interior of the vesicles___*becomes acidic (accumulates protons)*_____.

b. When electron transport is blocked in intact mitochondria, and an artificial pH gradient is imposed by lowering the pH of the medium, the synthesis of ATP_____*continues.*

c. FCCP is a drug that partitions itself into membranes and opens a channel specific for the passage of protons. When FCCP is added to a suspension of mitochondria, electron transport_____*continues*_____, and ATP synthes_____*stops*_____.

d. Under anaerobic conditions, the pH gradient _____*decreases*_____relative to aerobic conditions.

Chapter Six

Short Answer

1. Significance

a. the switch from H_2S to H_2O as a source of electrons for photosynthesis: *required that the photosynthetic process generate a very strong oxidizing agent using the energy of light. Photosynthetic organisms with this ability could exploit many more habitats. Their evolutionary success raised the level of O_2 in the atmosphere, paving the way for the evolution of organisms with aerobic capacity.*

b. antenna: *increase the variety of wavelengths that can be exploited for photosynthesis, and also the efficiency of photosynthesis, by increasing the number of electrons that can be transferred by the reaction center pigment per unit of time.*

c. fluidity of the thylakoid membranes: *high fluidity increases the rate of diffusion of the mobile electron carriers plastoquinone and plastocyanin between components of the photosynthetic apparatus.*

d. carotenoids: *most importantly, protect the photosynthetic apparatus from oxidation by reactive species of oxygen; they also absorb the blue and green wavelengths of light for photosynthesis.*

e. two photosystems, PSI and PSII: *essential for boosting the energy of electrons from water to a level sufficient to reduce $NADP^+$. Neither photosystem can alone provide sufficient energy to accomplish this highly endergonic task.*

f. alternate routes that electrons can take from ferredoxin (other than to NADPH): *provide a mechanism for photoautotrophs to reduce highly oxidized species, such as nitrate and sulfate, converting inorganic wastes into reduced molecules that can be incorporated into cellular macromolecules.*

g. thylakoid membranes are impermeable to protons: *prevents the proton gradient established by photosynthesis from dissipating by diffusion back into the lumen of the thylakoid.*

h. Rubisco binds to RuBP, not CO_2 or O_2: *leaves the substrate RuBP open to attack by either CO_2 or O_2, only one of which, CO_2 , results in productive carbon fixation. This lowers the efficiency of photosynthesis. Most enzymes exhibit high specificity for the most appropriate substrates. The absence of a binding site for CO_2 prevents the enzyme from expressing this level of specificity.*

i. peroxisomes are closely associated with chloroplasts: *reflects the interdependence of these organelles in the metabolism of glycolate.*

j. C_4 and CAM pathways: *provide certain plants a means to insulate their Rubisco from O_2, minimizing photorespiration and increasing the carbon-fixing efficiency of the plants.*

2. Compare and Contrast:
   a. chemoautotrophs vs. photoautotrophs
   Both types of organisms can reduce CO_2 and incorporate the carbon into organic molecules. Chemoautotrophs use the oxidation potential of simple inorganic molecules such as ammonia, hydrogen sulfide or nitrites as a source of energy, whereas photoautotrophs use the radiant energy of sunlight to accomplish this task.
   b. grana thylakoids vs. stroma thylakoids
   Together these form the internal membrane system of the chloroplast. Grana thylakoids are stacked pouches, whereas stromal thylakoids are flattened pouches that connect the grana. Many of the protein complexes involved in photosynthesis, including the cytochrome b_6f complex and PSII, are found in both types of membranes. Stromal thylakoids and the nonappressed membranes of the grana thylakoids are rich in PS I, ATP synthase molecules, and the loosely affiliated Rubisco molecules. There are differences in lipid composition as well. The stromal membranes and the nonappressed grana membranes have about equimolar amounts of MGDG and DGDG, whereas the ratio of MGDG to DGDG in appressed membranes is about three.
   c. light-dependent vs. light-independent reactions
   Both of these pathways are necessary steps in the de novo synthesis of organic compounds from CO_2. Light-dependent reactions include photolysis, reduction of $NADP^+$, and generation of a proton gradient that is used in the production of ATP. Light-independent reactions include the Calvin cycle– the pathway in which the energy of ATP and NADPH is used to fix carbon into organic compounds.
   d. cyclic vs. noncyclic photophosphorylation
   In noncyclic photophosphorylation electrons move from H_2O to $NADP^+$, generating a proton gradient that is used to synthesize ATP. The process requires two photosystems, PSII and PSI, for a double boost of light energy. In cyclic photophosphorylation, electrons removed from chlorophyll are replaced not by electrons derived from water, but by the same electrons that were removed– electrons move in a cycle from chlorophyll back to chlorophyll. Only one photosystem, PSI, is required. The energy that boosts the electrons from chlorophyll at the start is used to generate a proton gradient (just as in the noncyclic pathway) which is again used to make ATP.
   e. mesophyll cells vs. bundle sheath cells
   These cell types are characteristic of C_4 plants that spatially separate the processes of ATP synthesis and $NADP^+$ reduction from sugar production. The cell types are arranged in concentric cylinders, with mesophyll cells in the outer cylinder and bundle sheath cells forming the inner cylinder. ATP and NADPH are produced in the mesophyll cells, where CO_2 is fixed into the four-carbon compound, malate. Malate is transported into the bundle sheath cells where the CO_2 is released from malate, and serves as a substrate for Rubisco and eventual fixation into sugar. The inner cell layer is shielded from atmospheric O_2, and photorespiration is minimized.

Multiple Choice

1. c	2. b	3. a	4. d
5. a	6. b	7. c	8. a
9. a	10. d	11. a	12. a
13. c	14. c	15. d	16. b
17. b	18. a	19. c	20. b

Problems and Essays

1. a. *For H_2O:*

$$\Delta G^{o\prime} = -nF\Delta E_o$$

$$= -(2)(23.063\ kcal/V \cdot mol^{-1})(-0.320 - 0.816\ V)$$

$$= 52.4\ kcal/mole$$

For H_2S:

$$\Delta G^{o\prime} = -nF\Delta E_o^{\prime}$$

$$= -(2)(23,063\ kcal/V \bullet mol^{-1})(-0.320 - (0.25)\ V)$$

$$= 3.23\ kcal/mole$$

b. *At 680 nm:*

$$E = \frac{(1.58 \times 10^{-34}\ cal \cdot sec)(3.0 \times 10^8\ m/sec)}{680\ nm} \times (10^9\ nm/m)$$

$$= 6.97 \times 10^{-20}\ cal/photon \times (6.02 \times 10^{23}\ photons/mole)$$

$$= 4.2 \times 10^4\ cal/mole\ or\ 42.0\ kcal/mole$$

At 485 nm:

$$E = \frac{(1.58 \times 10^{-34}\ cal \cdot sec)(3.0 \times 10^8\ m/sec)}{485\ nm} \times (10^9\ nm/m)$$

$$= 9.77 \times 10^{-20}\ cal/photon \times (6.02 \times 10^{23}\ photons/mole)$$

$$= 5.88 \times 10^4\ cal/mole\ or\ 58.8\ kcal/mole$$

c. *At 680 nm, the 42 kcal is below the 54.2 kcal required for $NADP^+$ oxidize one mole H_2O. However the energy at 485 nm, 58.8 kcal, is sufficient for $NADP^+$ to oxidize H_2O.*

d.

$$\frac{52.4\ kcal/mole\ NADP^+\ reduced}{42.0\ kcal/mole\ photons} = 1.25\ mole\ photons/mole\ NADP$$

2. *For each six-carbon sugar, 18 ATPs are consumed, and 12 NADPHs are oxidized. Thus the ratio is three ATP to two NADPH. Plants rely on cyclic photophosphorylation for the generation of a proton gradient to synthesize ATP. $NADP^+$ is not reduced to NADPH during cyclic photophosphorylation.*

3.   *RL*   photolysis

             PT QH_2

             PT   cytochrome b_6f complex

             SC   ferredoxin-$NADP^+$ reductase

4. a. *C_4 and CAM plants;* b. *C_3 plants;* c. *CAM plants;* d. *C_4 plants*

5. a. *PSI: carotenes → Chl b → Chl a 660 → Chl a 670 → Chl a 678 → Chl a 685 → Chl a 690 → Chl a 700: Chlorophyll a 700 is acting as the reaction center in this photosynthetic unit.*

PSII: xanthophylls → Chl b → Chl a 660 → Chl a 670 → Chl a 678 → Chl a 680 → Chl a 685: Chlorophyll a 685 is acting as the reaction center in this photosynthetic unit.

   b. Light of 700 nm wavelength is absorbed by the reaction-center chlorophyll of PSI, but by itself, PSI cannot split water, and 700 nm light does not have enough energy to activate PSII. Without a source of electrons, noncyclic photosynthesis ceases. Adding a beam at 680 activates PSII, which can then split water and send electrons to PSI. Noncyclic photosynthesis occurs.

6.   (1) The thylakoid membranes serve as a scaffolding for the photosystems I and II, and the electron carriers. The membranes keep the components of the photosystems in the proper orientation such that the energy of light is funneled into the photoreaction center, and the electrons are passed from donor to acceptor in the proper sequence. (2) The thylakoid membranes separate protons such that a gradient can be established and exploited in the synthesis of ATP. The chloroplast ATP synthase is held in place with the proton channel CF_O open to the proton-rich lumen and the catalytic CF_1 oriented in the stroma for ATP production. (3) The inner and two outer chloroplast membranes separate the reactions of carbon fixation from the catalytic reactions of the cytoplasm. These two pathways share many intermediates, and intermixing of the two would result in breakdown of the intermediates before they could be incorporated into sugars.

Chapter Seven

Short Answer

1.   Significance
   - a. glycocalyx: *mediates cell-to-cell interactions, provides mechanical protection to the cell, and serves as a barrier to particles reaching the plasma membrane.*
   - b. basement membrane: *may maintain the polarity of the cell, determine the path of cellular migration, separate adjacent tissues, and act as a barrier to the passage of macromolecules.*
   - c. integrins: *attach cells to elements of their microenvironment, including basement membranes and artificial substrates (in cultured cells), and, in some cases to other cells (via binding to IgSF proteins on adjacent cells).*
   - d. RGD sequence: *is recognized as a ligand by many integrins. This sequence occurs on several important ECM proteins, including fibronectin, laminin and collagen, as well as on other extracellular proteins.*
   - e. hemidesmosomes: *anchor cells to basement membranes. These cell-to-cell connections are especially important in epithelial tissues, such as the epidermis.*
   - f. cadherins: *serve as the major protein connecting like cells of a tissue to one another.*
   - g. tight junctions: *prevent leakage between cells of a tissue. This is especially important in cells that line body cavities and must maintain different environments on either side of the lining.*
   - h. gap junctions: *allow communication (GJIC) between adjacent cells. This is especially important in tissue cells that have to act as a coordinated unit, such as those of the heart.*
   - i. plasmodesmata: *permit intercellular communication and the exhange of small-molecular-weight solutes between adjacent cells of plant tissues.*
   - j. pectins: *form extensive hydrated gels that act as molecular sieves determining the size of solutes that can penetrate plant cell walls and reach the plasma membrane. Pectins can also fragment when the plant is attacked by pathogens, stimulating a defensive response. In addition, pectins cement the cell walls of adjacent plants together.*

2.   Compare and Contrast:
   - a. glycocalyx vs. extracellular matrix
     These are both extracellular structures of cellular origin. The glycocalyx is closely affiliated with the cell membrane; it comprises carbohydrate chains of both glycoproteins and glycolipids of the cell membrane. Some cell types have secreted extracellular materials as part of the glycocalyx. The extracellular matrix is a network of extracellular materials that lies beyond the immediate

vicinity of the cell membrane. In some tissues, such as bone and cartilage, the ECM is the dominant feature.

b. fibronectin vs. integrin

These proteins both participate in cell adhesion, but one is an extracellular protein of the ECM (fibronectin), and the other is an integral membrane protein (integrin). Fibronectin has binding sites for other components of the ECM and for proteins of the cell surface, including integrin. Integrins are membrane-spanning heterodimers with binding sites for fibronectin, other RGD-containing proteins, and other proteins of the ECM. In some cells, fibronectin can act as an integrin-specific ligand, and integrin, as a fibronectin receptor.

c. focal contacts vs. hemidesmosomes

These two types of cellular adhesive structures occur in different settings. Focal contacts bind cells to artificial substrates in vitro, whereas hemidesmosomes bind cells to basement membranes in vivo. There are structural differences, notably that focal contacts are characterized by intracellular actin filaments and hemidesmosomes are characterized by intracellular keratin filaments.

d. tight junctions vs. gap junctions

These two types of cell-to-cell connections differ in both structure and function. Tight junctions form "belts" that surround the periphery of the cell and prevent leakage of solutes between cells of a tissue. They are most prevalent in epithelial tissues defining a body compartment and maintaining compositional differences between compartments. Gap junctions connect adjacent cells and permit the diffusion of small-molecular-weight substance between cells. They are most prevalent in tissues that must act in synchrony, such as cardiac muscle and smooth muscle.

e. primary cell walls vs. secondary cell walls

These two types of plant cell walls characterize plant cells at different stages of maturity. Primary cell walls surround growing plant cells and allow some extensibility accommodating further growth. Secondary cell walls occur in more mature plant cells, and are more rigid. They contain more cellulose and more than primary walls. Secondary walls are superior support structures.

Multiple Choice

1. c	2. b	3. d	4. a
5. a	6. d	7. b	8. c
9. d	10. b	11. d	12. a
13. b	14. c	15 c	16. a
17. a	18. b	19. c	20. b

Problems and Essays

1. One way to determine which ECM protein is acting as an attractant for growing axons is to coat the bottom of a petri dish with alternating strips of collagen and laminin. Introduce cultured embryonic nerve cells and watch for outgrowth from the cell bodies. This experiment was done by R. W. Gunderson (1987. Devel. Biol. 121:423) and the axon outgrowth was restricted to the areas of the dish coated with laminin. Other experiments might be equally convincing.

2.

   _**EX**__ laminin

   _**EX**__ collagen

   _**MS**__ heparan sulfate proteoglycans (HSPG)

   _**MS**__ connexin

   _**EX**__ fibronectin

___IN___ keratins affiliated with hemidesmosomes

___MS___ integrin

___MS___ selectin

___MS___ NCAM

___IN___ catenins

___IN___ actin

3. *(1) The lower layer of the epidermis attaches to its basement membrane via hemidesmosomes that involve integrin-RGD binding. Artificial RGD peptides would compete for these binding sites on integrin. (2) Some cell-to-cell adhesions, particulary between integrins and NCAMs or VCAMs, involve integrin-RGD binding. Artificial peptides would compete for these binding sites. (3) Integrin binding onto epithelial cells of the venules plays a role in leukocyte extravasation during inflammation. This is another potential site of side effects from artificial RGD peptides.*

4.

5.

cell surface protein	expression increased or decreased in cancer cell	contribution to cell's malignant character
laminin receptors	*increased*	*Increases binding of metastatic cells to laminin in basement membranes of blood vessel walls; increases ability of the cell to invade tissues.*
fibronectin receptors	*decreased*	*Cells can escape from their tissue of origin by sliding over ECM, without sticking.*
integrins	*decreased*	*Because integrins act as fibronectin receptors, the contributions listed above apply here, also; i.e., cells escape from tissues by sliding over the ECM.*
cadherins	*decreased*	*Cells do not recognize or bind like types and can easily break away from their tissue of origin.*
gap junctions	*decreased*	*GJIC is reduced; cell growth is not tightly regulated.*

6. a. *Gap junctions facilitate intercellular communication enabling the tissue to act as a unit.*
    b. *Tight junctions prevent secreted enzymes from diffusing back into tissues of the pancreas.*
    c. *Desmosomes give this tissue its strength and ability to be stretched. Some gap junctions may coordinate contractions during labor.*
    d. *Tight junctions prevent reverse diffusion of salts that are pumped against a gradient.*

e. *Gap junctions keep rings of contractile tissue coordinated so that contractions are coordinated and productive.*

Chapter Eight

Short Answer
1. Significance
   a. regulated secretion: *one way in which a cell responds to specific stimuli, either from the environment or from within the cell.*
   b. cell fractionation: *a technique used for in vitro study whereby subcellular fractions or structures are isolaated based on differences in physical properties.*
   c. signal sequence: *targets the nascent protein with this stretch of amino acids (the signal sequence) for a specific intracellular or secretory destination.*
   d. trans Golgi network: *serves as the site for sorting different vesicles (and their protein cargos) that arise from the Golgi.*
   e. retrograde movement of Golgi vesicles: *allows the contents of Golgi vesicles to be recyled. As Golgi cisternae move from cis to trans, some enzymes and other components can be moved back to the cis surface cisternae.*
   f. phosphorylated mannose residues on TGN glycoproteins: *target trans Golgi network (TNG) proteins for the lysosomes.*
   g. lysosomes: act as cellular digestive organelles. *Lysosomes digest both exogenous particles that are taken into the cell by phagocytosis and cellular organelles that are damaged or aged.*
   h. acid hydrolases: *the digestive machinery of lysosomes, these enzymes have low pH optima and catalyze the hydrolytic breakdown of all types of biologic macromolecules.*
   i. receptor-mediated endocytosis (RME): *bulk-uptake mechanism in response to specific extracellular solutes or particles.*
   j. coated pits: *result from receptor-mediated endocytosis and form the cell surface precursor to endocytic vesicles, with specific cargo being brought into the cell.*

2. Compare and Contrast:
   a. pulse vs. chase
      These two techniques are used together as a method of tracing the incorporation of precursors into the macromolecules of various cellular organelles. During the pulse phase, cells are given radiolabeled building-block molecules, such as amino acids. After a brief time period, the excess radiolabeled precursors are washed away, and those that have been incorporated into macromolecules are "chased" by measuring which cellular fractions or organelles have become radiolabeled.
   b. protein synthesis on membrane-bound vs. free ribosomes
      Proteins that are synthesized on ribosomes affiliated with the endoplasmic reticular membrane (rough ER) contain specific signal sequences that bind to signal recognition particles (SRPs) recognized by SRP receptors on the ER. These proteins are destined for secretion, specific membrane-bond organelles, or service as integral membrane proteins. Proteins lacking the signal sequence are synthesized on free ribosomes, and comprise a group destined for the cytosol, the nucleus, the mitochondria or chloroplasts.
   c. synthesis of *N*-linked vs. *O*-linked oligosaccharides
      The saccharide groups that are added to proteins synthesized on the rough ER can be connected in two ways: either via the nitrogen atom of an asparagine residue (N-linked), or via the oxygen atom of a serine or threonine residue (O-linked). The initial steps in the synthesis of N-linked glycoproteins occurs in the rough ER, where a core segment of the carbohydrate chain is transferred from a lipid carrier, dolichol phosphate, to the nascent peptide. The core sequence of sugars is modified and elaborated in the Golgi. O-linked carbohydrates are added to proteins entirely in the Golgi.

d. autophagy vs. phagocytosis

   Both of these processes involve intracellular digestion. In phagocytosis, exogenous particles are taken into the cell in endocytic vesicles which fuse with endosomes and lysosomes where digestion occurs. In autophagy, old or damaged organelles are surrounded by membrane derived from the endoplasmic reticulum, and the resulting vesicle fuses with a lysosome initiating digestion.

e. LDL vs. HDL

   These are particulate protein-lipid complexes that travel via the bloodstream, transporting cholesterol between organs. Low-density lipoproteins (LDLs) carry cholesterol from the liver (which gets it from the digestive tract, where it is absorbed from food) to the cells of the tissues. High-density lipoproteins carry excess cholesterol from the tissues back to the liver where it is secreted as part of the bile. Both are taken into cells via receptor-mediated endocytosis. High blood levels of LDL are correlated with atherosclerosis and heart disease. High blood levels of HDL are associated with a decreased risk of heart disease.

Multiple Choice

1.	b	2.	a	3.	d	4.	c
5.	b	6.	d	7.	b	8.	b
9.	c	10.	a	11.	d	12.	b
13.	a	14.	b	15.	d	16.	c
17.	a	18.	a	19.	a	20.	b

Problems and Essays

1. *RER*   fibronectin

    FR   cytoskeletal proteins such as actin

    RER   acid hydrolases

    RER   collagen

    FR   triskelion

    FR   the enzymes of glycolysis

    RER   adaptin

    RER   LDL receptor

    RER   signal peptidase

    RER   protein disulfide isomerase

    FR   mitochondrial proteins coded for by nuclear DNA

2. *The bacterial cell is engulfed by the macrophage and radiolabel first appears in the cell in a large endocytic vesicle. The endocytic vesicle then fuses with an early endosome, near the cell periphery. Gradually, as the endosome moves toward the cell interior, the radiolabel has more internal presence. Lysosomes fuse with the late endosome, and the proteins of the bacterium are digested. Radiolabeled amino acids are exported from the late endosome and incorporated into cellular proteins of all types.*

3. a. *The proteolytic cleavage of proinsulin to insulin occurs during the maturation of secretory vesicles, after they leave the trans Golgi network.*

    b.

	anti P	anti I	anti clathrin
mature secretory vesicles	-	+	-
trans Golgi network	+	-	+
cis and medial Golgi cisternae	+	-	-

4. The C- and N-termini must be directed toward the lumen of the ER, the Golgi, and the secretory vesicles. The side of the membrane facing the cytoplasm will always face the cytoplasm.

5.

Experimental Conditions	completely synthesized?	signal sequence?	translocated across a membrane?
a. mRNA codes for a cytosolic protein; no SRP present; no SRP receptor present:; no microsomes present	yes	no	no
b. mRNA codes for a secretory protein; no SRP present; no SRP receptor present; no microsomes present	yes	yes	no
c. mRNA codes for a secretory protein; exogenous SRP added; no SRP receptor present; no microsomes present	no	no	no
d. mRNA codes for a secretory protein; exogenous SRP added; free (not membrane-bound) SPR receptor added; no microsomes	yes	yes	no
e. mRNA codes for a secretory protein; exogenous SRP added; exogenous SRP receptor added; microsomes added	yes	no	yes

6. Ferrotransferrin binds to the transferrin receptor on cell membranes, and a coated pit is formed. A clathrin-coated endocytic vesicle enters the cell, containing a membrane-bound transferrin receptor-ferrotransferrin complex. Clathrin dissociates from the vesicle, which then fuses with an early endosome. As lysosomes bind to the endosome, the pH of the vesicle decreases, and iron diffuses away from the ferrotransferrin. The ferrotransferrin, however, stays bound to the membrane-bound receptor. Iron is transported out of the late endosome and into the cytosol, where it is used in the synthesis of iron-affiliated proteins. The late endosome migrates to the cell surface and fuses with the plasma membrane, where the transferrin-transferrin receptor complex becomes part of the plasma membrane. Transferrin dissociates from the transferrin receptor at the neutral pH of the cell exterior and enters circulation, where it can participate in another round of iron uptake.

Chapter Nine

Short Answer
1. Significance
   a. overlapping functions of cytoskeletal proteins: *provide redundancy that protects the cell from the consequences of a defective protein.*
   b. microtubule-organizing centers MTOC: *act as sites of origin for microtubule assembly.*
   c. the high metabolic cost of microtubule polymerization: *allows a cell to independently control the rates of assembly and disassembly. Tubulin dimers bound to GTP at the end of a microtubule favor polymerization, whereas those bound to GDP have less affinity for each other and are more likely to disassemble.*
   d. γ–tubulin: *may serve as the nucleating tubulin upon which each new microtubule is assembled.*
   e. intermediate filaments occur only in animal cells: *may reflect the absence of cell walls in animal cells. Cells from all other kingdoms are protected and supported by cell walls, a job that is done by intermediate filaments in animal cells.*
   f. high degree of conservation in the structure of actin proteins: *evidence that virtually every amino acid participates in some essential activity of the protein. This is a reflection of actin's many roles in cells.*
   g. experiments of A. Huxley, Niedergerke, H. Huxley and Hanson: *provided the first evidence that the shortening of a skeletal muscle cell during contraction is due not to the*

shortening of individual proteins, but rather to the sliding of interdigitated filaments past one another in a regulated manner.

h. actin-binding proteins: *determine the organization, assembly, physical properties, and interactions of actin molecules within cells.*

i. contractile stress fibers: *may maintain tension between a cell and its substrate by isometric contraction, the generation of tension without a change in length.*

j. contact inhibition of movement: *is partially responsible for the aggregation of cells, particularly epthelial cells, into tissues.*

2. Compare and Contrast:

a. microtubules vs. microfilaments
    These are two types of cytoskeletal elements that occur in virtually all types of eukaryotic cells. Both play important roles in cell motility, cell division, and the maintenance of cell shape. Both, too, can be in a state of dynamic instability in cells, with relative rates of assembly and disassembly reflecting the needs of the cell at any given time. They differ in size, occurrence, the motor proteins with which they affiliate, the subunits of which they are composed, the manner in which they are regulated, and the specific roles they play within cells.

b. dynein vs. kinesin
    Both dynein and kinesin are large motor proteins that convert the chemical energy of ATP into movement. Both are found affiliated with microtubules, although only dynein occurs on the microtubules of cilia and flagella. Kinesin is a plus-end directed microtubular motor, and dynein, among its other roles, is a minus-end directed microtubular motor. In spite of their similarities in function, they are not homologous proteins, and they assume quite different three-dimensional shapes. They are not members of a protein family.

c. basal bodies vs. centrioles
    Both of these intracellular structures serve as microtubule-organizing centers in cells, and, along with their affiliated proteins, are the sites of microtubular nucleation. They share identical structures of nine evenly spaced fibrils each comprised of three microtubules. Centrioles occur near the nuclei of some but not all eukaryotic cells. During interphase, they are the sites of microtubular convergence, probably a reflection of their roles in initiation. During cell division, the mitotic spindle originates from the centrioles. Basal bodies occur near the base of cilia and flagella, where they give rise to the microtubular axoneme.

d. muscle fibers vs. myofibrils
    The term muscle fiber is used to refer to a single skeletal muscle cell, but each cell is actually the multinucleated product of fusion of many embryonic myoblasts. Myofibrils are thinner cylindrical strands, hundreds of which may occur in a single muscle fiber. Myofibrils are composed of a linear array of sarcomeres.

e. pseudopodia vs. lamellipodia
    Single cells have the ability to "crawl" over their respective substrates by extending cytoplasmic projections from the leading edge. Cells that move by amoeboid crawling extend broad, rounded protrusions called pseudopodia. Cytoplasm streams into these protrusions from the interior of the cell affecting forward movement. Cells that are normally part of a tissue, such as cultures fibroblasts, send out broad, flattened, sheetlike projections called lamellipodia. Lamellipodia form temporary adhesions with the substrate over which the cell crawls. Both types of cytoplasmic projections involve actin.

Multiple Choice

1. c	2. c	3. a	4. d
5. d	6. d	7. b	8. a
9. c	10. a	11. b	12. a
13. a	14. b	15. d	16. a
17. a	18. b	19. a	20. d

Problems and Essays

1. a. *The negative slopes of these lines are greater than the positve slopes, indicating that disassembly is more rapid than assembly.*
   b. *Line A probably represents activity at the plus end. In general, the plus end is more active than the minus end, and both assembly and disassembly at the plus end are more rapid. The slopes of the A line in both positive and negative directions are greater than those of the B line.*
   c. *dynamic instability*
   d. *Changes in either Ca^{2+} or cAMP levels have been shown to influence the dynamic instability of microtubules.*
   e. *Assuming that the microtubule you have been watching is not involved in a spindle, it would probably disassemble entirely, and the tubulin subunits would be used to build a spindle.*

2.
__AMP-PNP, colchicine__	axonal transport
__colchicine__	mitotic spindle formation
__taxol__	mitotic spindle disassembly
__AMP-PNP, cytochalasin__	acrosomal reaction
__AMP-PNP, cytochalasin__	phagocytosis
__AMP-PNP, cytochalasin__	cytokinesis

3. a. *The presence of both types of filaments indicates that this is a cross section at the A band, but not in the region of the H zone.*
   b.

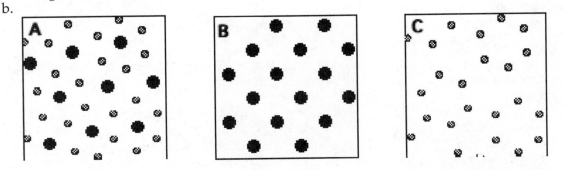

   c. *It might be possible to see some engaged cross-bridges in contracted muscle; these are not visible in relaxed muscle. They would appear as spokes radiating from the thick filaments and contacting the thin filaments.*

4. *The microtubules that form the mitotic spindle are in a state of dynamic assembly and disassembly, both of which are necessary for proper function. The alignment of chromosomes at the equator during metaphase, for example, occurs by the lengthening and shortening of the spindle microtubules. Most importantly, the separation of chromosomes during anaphase results from the shortening of the spindle fibers, a process that is inhibited by taxol.*

5. Without either the nexin links between adjacent microtubules or the radial spokes that constrain their movements, the ATP-driven action of dynein slides the fibrils relative to one another until they telescope outward from their original positions. It is possible for each fibril to slide completely off both of its adjacent fibrils until they are lined up end to end. If this happens, the length of the axoneme will be equal to the length of all the fibrils in the ring, or nine times its original length.

6.

_____*NONE*_____	tropomyosin
_____*PM*_____	acetylcholine receptor
_____*SR*_____	Ca^{2+}-ATPase
_____*NONE*_____	titin
_____*SR*_____	Ca^{2+}-release channels

7.

	Ciliary Movement	Skeletal Muscle Movement
Filament Type	*microtubule*	*microfilaments and myosin*
Motor Protein	*dynein and kinesin*	*myosin*
Fuel Molecule	*ATP*	*ATP*
Name of the Functional Unit of Movement	*axoneme*	*sarcomere*

Chapter Ten

Short Answer

1. Significance
   a. crossing over: *one source of new combinations of genes in gametes, and hence in offspring. Crossing over is responsible for incomplete linkage– not all traits that occur on the same parental chromosome are passed on together to offspring.*
   b. polytene chromosomes: *from the point of view of the insect, these amplified chromosomes provide the many gene copies that are required to produce large amounts of proteins efficiently. For researchers, the bands on these giant structures correlate with specific genes, providing visual confirmation of the genetic maps constructed using crossover frequencies.*
   c. topoisomerases: *catalyze the essential first step in DNA replication or transcription– unwinding supercoiled DNA.*
   d. T. H. Morgan's experiments with *Drosophila*: *established that linkage is incomplete, and that the reason for this is that pieces of homologous pairs of chromosome cross over. Morgan used crossover frequencies to construct the first genetic map.*
   e. DNA supercoiling: *packages long stretches of DNA into small, compact units.*
   f. Barbara McClintock's experiments with maize: *the first to show that the genome of a eukaryote is not immutable. McClintock established that some genetic units, which she named transposable elements, can move within the genome.*

g. single nucleiotide polymorphisms (SNPs): *sites in a genome that are both abundant and highly polymorphic and make excellent genetic markers. These single nucleotide sites may also give insights into individual susceptibility to certain complex diseases, as well as the how an individual might react to certain drugs or other treatments.*

h. Avery, Macleod and McCarty's experiments with pneumococcus: *the first to establish that the "transforming principle," the substance that could make a nonvirulent pneumococcus become virulent, is nucleic acid. By inference, it was determined that the genetic material is DNA.*

2. Compare and Contrast:

a. gene vs. allele
A gene is that segment of DNA that codes for a protein. Genes occur at particular positions on chromosomes, called loci. An allele is the exact form of a gene at a given locus. One gene may have several different alleles, all residing at the same locus on either different members of homologous pairs or chromosomes from different individuals. Thus, while any one individual cannot have more than two alleles for a gene, a group of individuals may harbor many different alleles for that gene.

b. Mendel's law of segregation vs. Mendel's law of independent assortment
The law of segregation refers to the manner in which members of homologous pairs of chromosomes, and the genes that reside on them, segregate. One member of each pair, or, as Mendel saw it, one form of each trait, ends up in each gamete. Independent assortment says that the manner in which members of any one pair of chromosomes segregate is independent of how the members of any other pair segregate, or, as Mendel saw it, different traits move into the next generation independently of each other.

c. genomic complexity of prokaryotes vs. genomic complexity of eukaryotes
Prokaryotes have one kind or class of DNA: nonrepeating, coding DNA. Eukaryotes have three general classes of DNA, distinguished by the degree of repetition: nonrepeating DNA, moderately repetitive DNA, and highly repetitive DNA. The differences are best illustrated using reannealing curves. Prokaryotes have simple sigmoidal curves, whereas eukaryotic curves have three plateaus corresponding to the three classes of DNA.

d. number of human protein coding sequences vs. number of human proteins
The human genome is estimated to have about 30,000 protein-coding sequences, but 60% of those are believed to undergo alternate splicing or posttranslational modifications, resulting in more than one protein for these coding sequences. Thus the number of human proteins is far greater than the number of protein coding sequences in the genome.

e. conserved vs. variable sequences
Conserved sequences are nucleotide sequences that are similar among closely related species; variable sequences are much less similar. Parts of genomes that exhibit a great deal of conservation among species represent important genes or regulatory sequences—those that have not undergone much change over evolutionary time. Stretches of nucleotide that vary considerable among species appear to have undergone random changes with little impact on survival or reproduction, and are thus considered less important.

Multiple Choice

1.	b	2.	a	3.	c	4.	d
5.	c	6.	d	7.	c	8.	d
9.	c	10.	a	11.	c	12.	a
13.	d	14.	b	15.	d	16.	d
17.	d	18.	a	19.	c	20.	b

Problems and Essays

1.

$$1.16 \; ribosomes \,/\, sec \cdot 60 \, sec \,/\, min \cdot 60 \, min \,/\, hr \cdot 24 \; hr = 100{,}224 \; ribosomes \,/\, 24\,hr$$

$$100{,}224 \; ribosomes \,/\, 24 \; hr \,/\, gene \times x \; genes = 10 \times 10^{6} \, ribosomes$$

$$x = 99.78 \; genes$$

In fact, there are between 200 and 2000 copies of the genes for the various ribosomal RNAs in the human genome.

2. The first Alu sequence probably arose by a duplication of the gene for the 7SL RNA, an event that most likely occurred sometime early in the evolution of mammals. Since Alu has the features of a transposable element (i.e., terminal inverted repeats), the duplication mechanism may have involved reverse transcription of the 7SL RNA, followed by insertion of the new DNA at a different place in the genome. Over time, the Alu sequence underwent many such duplications, and the nucleotide sequences of the duplicated DNA changed by mutation, eventually becoming a very large "family" of SINEs.

3.

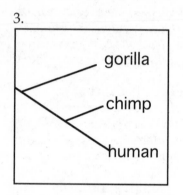

a. Comparisons of the sequence similarities between humans, chimps and gorillas shows that humans and chimps are 80% identical and chimps and gorillas are also 80% identical. Human and gorillas, however, are only 60% identical. Based on this single sequence, we conclude that humans are more closely related to chimps than to gorillas, and gorillas and chimps are equally distantly related as are humans and chimps.

b.   Based on this sequence, the gorilla is most closely related to the common ancestor with only three nucleotide differences (85% identical). The chimpanzee is 62% identical, and the human is 50% identical. The sequences that most resemble those of the common ancestor are usually found in the more primative species. Thus the answer to part a does not change with this new information. Keep in mind, however, that this is one short segment of a single gene. To establish phylogenies with certainty requires many such analyses from many different regions of the genome.

4.   a. $\dfrac{30{,}000 \; genes}{23 \; chromosomes} = 1304 \, genes \,/\, chromosome$

b. $\dfrac{3 \times 10^{9} \; base \; pairs}{3 \times 10^{4} \; genes} = 100{,}000 \; base \; pairs \,/\, gene$

c. $\dfrac{(3 \times 10^{9} \; base \; pairs) \times 0.05}{3 \times 10^{4} \; genes} = 5000 \; base \; pairs \,/\, gene$

5.

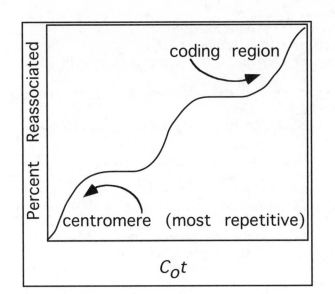

The most repetitive DNA, much of which resides near the centromeres, reanneals first because there are so many copies of it that each strand rapidly finds a complementary strand. The coding regions are usually the least repetitive and reanneal last. These regions are slow to reanneal because each single strand must find its one and only complementary strand.

6.   While on the same chromosome, these two genes are far enough apart that they are separated by crossover events at a very high frequency; they appear to act independently. This diagram shows one possible fluorescence pattern that might result from the FISH technique where the distance between the two genes on the chromosome is maximized. Other patterns are possible.

7.   Band 3 is the only band that differs in a consistent way between those with and those without the disease. This consistency means that the DNA fragment is probably close to, if not part of, the gene for the disease. Band 1 is highly conserved, and band 2 is highly polymorphic between individuals.

Chapter Eleven

Short Answer

1.   Significance
   a.   mRNA as an intermediary between DNA and protein: *This indirect system of protein synthesis enables amplification of the amount of protein that can be produced from any one gene. One gene can be transcribed many times, and each RNA can be used many times as a template for protein synthesis. This may evidence to suggest that the first macromolecules in cellular evolution were RNA.*
   b.   the highly exergonic nature of RNA polymerization: *decreases the likelihood that RNA synthesis can proceed in the reverse direction by the same mechanism as polymerization.*
   c.   slight differences in the sequences of nucleotides in promoter regions: *deviations from the "consensus sequence" change the binding affinity for the RNA polymerase, and hence the likelihood that a gene will be transcribed. This may be a mechanism for regulating which genes are transcribed, and how often.*

d. methylguanosine cap: *prevents digestion of the 5' end by nucleases, aids in the transport of mRNA out of the nucleus, and plays a role in initiating translation.*

e. alternate splicing: *the same gene can code for several different polypeptides.*

f. general transcription factors: *fine tune and regulate the transcription of genes by bringing together the initiation complex.*

g. Shine-Dalgarno sequence: *In prokaryotic cells, this sequence orients the mRNA on the small ribosomal subunit and identifies the initial AUG codon.*

h. our changing concept of the gene: *reflects our ever changing understanding of the complexity of information transfer in cells.*

2. Compare and Contrast:

a. "one gene – one enzyme" vs. "one gene – one transcription unit"
These phrases reflect our understanding of the relationship between genes and traits, at two different times in the history of cell biology. The first phrase, derived from the work of Beadle and Tatum, described the finding that one gene carried the information to construct one enzyme. The second phrase expanded Beadle and Tatum's understanding, describing the observation that genes carry elements other than just the information for constructing an enzyme These elemnts include promoter sequences, introns, etc.

b. higher-order structure of proteins vs. higher-order structure of RNA
Secondary and tertiary structure are essential to the roles of both proteins and RNA. In proteins, higher-order structure is mainly a result of hydrophobic amino acids that interact better with each other than with the aqueous medium, hydrogen bonding that stabilizes helices and ß sheets, and the formation of covalent bonds between cysteine residues. In RNA, higher-order structure results primarily from complementary base pairing between nucleotides.

c. Pribnow box vs. TATA box
These are both recognition sites for RNA polymerases found on DNA. The Pribnow box occurs in prokaryotic genes, whereas the TATA box is a feature of eukaryotic genes. The sequence of bases in these two promoters is very similar. The Pribnow box most often has the sequence TATAAT, whereas the consensus sequence of the TATA box is TATAA.

d. upstream promoters vs. internal promoters
The promoter region on DNA is the sequence of bases that is recognized by the RNA polymerase prior to initiation of transcription. In most genes, the promoter region occurs upstream from the template region. However, for genes that are transcribed by RNA polymerase III, for example the genes for the 5S ribosomal RNA, the RNA polymerase binds downstream of the transcription initiation site, to a sequence within the coding DNA.

e. frameshift mutations vs. substitution mutations
A frameshift mutation results when one or two nucleotides are erroneously deleted from or added to a gene. When the gene is transcribed and translated, the reading frame of the message is shifted by one or two nucleotides, and all the subsequent triplet codons are misread. A substitution mutation occurs when one nucleotide base is substituted for another. The result can be harmful, but often it is inconsequential, especially if the substitution occurs in the third base of a triplet coding for an amino acid. Even if the substitution occurs in one of the first two bases, only one amino acid on the protein is altered; the ability of the protein to function may be unaffected.

Multiple Choice

1. b	2. c	3. b	4. a
5. d	6. c	7. b	8. c
9. d	10. b	11. a	12. b
13. d	14. a	15. a	16. c
17. d	18. a	19. d	20. d

Problems and Essays

1. The sequence TATAA occurs on the complementary strand rather than the actual template strand of the DNA. Recall that regions of DNA are named according to the RNA that is transcribed rather than the template. Recall also that the textbook specifically indicates that the TATA box has the sequence 5'-TATAAA-3'. The 5' and 3' are written at the beginning and the end of the sequence respectively– the way that RNA is synthesized. Prokaryotic chromosomes are circular and lack free ends, and eukaryotic genes do not occur at the ends of DNA strands, so the designations 3' and 5', which refer only to free ends, are not relevant to the DNA.

2.
nucleus	hnRNA	cytoplasm	tRNA
cytoplasm	mRNA	nucleus	pre-RNA
nucleus	snRNA	nucleus	nascent rRNA
cytoplasm	rRNA	nucleus	45S pre-rRNA

3. The cDNA is complementary to the mRNA, which is comprised only of the exons from the original gene. It is 5.5 kilobases, and retains only one of the EcoR1 restriction sites. Two fragments result from digestion of the cDNA with EcoR1, and they are 3.0 kb and 2.5 kb long.

4. a. The most abundant codons that result are:

CCC	CCU	UCC	CUC
PRO	PRO	SER	LEU

They code for these amino acids:

Other possible(but less likely) codons are:

UUC	CUU	UCU	UUU
PHE	LEU	SER	PHE

They code for these amino acids:

   b. Proline would be most abundant.

   c. Phenylalanine would be least abundant.

5.
The RNA polymerase needs 1 NTP per base x 3 bases per amino acid x 8 amino acids	= 24 NTP.
The aminoacyl tRNA synthetases use 1 ATP per amino acid x 8 amino acids	= 8 ATP.
Translocation requires 2 (+) GTP per amino acid x 8 amino acids	= 16 GTP(+).
Finally, release factor requires 1 GTP per peptide	1 GTP.
The total number of NTPs required is at least	49 NTP.

   Aside from the higher level of activity in children relative to adults, the process of growth involves extensive protein synthesis as a part of both cell division and cell growth. The energy demands of a growing child are higher per pound than that of an individual who has ceased growing.

6. The amino acid sequence in the protein:

MET-VAL-HIS-LYS-ARG-THR-LEU-VAL stop

was translated from the mRNA:

AUG GUU CAU AAA **AGA** ACC UUA GUU UAA

or some similar sequence.

The mutant protein:

MET-VAL-HIS-LYS-GLU-PRO

is truncated and must have resulted from a frameshift mutation that inserted a premature stop codon. One possible mutation could result in the following RNA:

AUG GUU CAU AAA *GAA CCU **UAG** UUU AAstop

where the asterisk represents the loss of an adenine. Other deletions might give the same mutant protein.

7.   *Using the enzyme polynucleotide phosphorylase and a mixture of all the nucleotide triphosphates, make a mixture of synthetic RNAs that differ in random ways in both their base compositions and lengths. Pour this mixture over a column of inert particles to which ATP has been covalently bound. Discard the RNAs that do not adhere to the column particles. Dislodge the RNAs that adhere by using a highly-concentrated salt solution. Using a combination of reverse transcriptase, followed by RNA polymerase and the appropriate precursors and cofactors, replicate the RNAs that adhere, and repeat the column selection step under more stringent conditions, e.g., in the presence of a mild salt solution that requires the RNAs to have a higher affinity for the ATP in order to stick to the column. Repeat this process until you have a population of RNAs with a very high affinity for ATP.*

Chapter Twelve

Short Answer

1.   Significance
   a.   nuclear pore complex: *regulates the movement of proteins and RNAs between the nucleus and the cytoplasm by recognizing and transporting these molecules in the appropriate direction.*
   b.   nuclear localization signal: *enables proteins synthesized in the cytoplasm to be recognized and transported through the nuclear pores into the nucleus.*
   c.   evolutionary conservation of histone proteins: *evidence that every amino acid in these proteins is functionally important.*
   d.   H1 histone: *maintains the 30 nm filament structure of chromatin. Without the H1 histone, the 30 nm filament uncoils to form "beads on a string" nucleosomes.*
   e.   nuclear matrix: *maintains the shape of the nucleus; organizes the loops of chromatin; guides RNAs to nuclear pores.*
   f.   selective gene amplification: *increases the number of copies of certain genes, particularly some ribosomal RNA genes, so that the expression of those genes is increased.*
   g.   DNA rearrangement of antibody genes: *provides one mechanism for the synthesis of a diverse variety of antibodies from a limited number of antibody genes.*
   h.   heterodimerization of transcription factors: *expands the diversity of transcription factors that can be generated from a limited number of peptides, and allows one protein to alter the DNA binding properties of another.*
   i.   DNase I hypersensitive sites: *identify genes in the process of active transcription.*
   j.   acetylated histones: *disrupt the structure of the nucleosome and partially dislodge the DNA from the histone proteins, making the DNA more likely to be transcribed.*

2.   Compare and Contrast:
   a.   heterochromatin vs. euchromatin
      These are forms of interphase chromatin and represent two different levels of condensation. Heterochromatin remains highly condensed and is not actively transcribed, whereas euchromatin is dispersed and can serve as a template for transcription.
   b.   constitutive vs. facultative heterochromatin
      These are both forms of highly condensed interphase chromatin. Constitutive heterochromatin remains condensed throughout the cell cycle and is never transcribed. It mostly consists of highly repetitive, noncoding DNA and is located around the centromere region of the chromosome. Facultative heterochromatin is condensed and transcriptionally silent during certain phases of the life cycle, but may be dispersed and transcribed during other phases. The inactive X chromosome in females, the Barr body, is an example of facultative heterochromatin.
   c.   repressible vs. inducible operons
      Operons are prokaryotic transcriptional units, including their regulatory sequences, that code for one or several proteins with related functions. An inducible operon is one whose structural genes will be transcribed in response to some key metabolic substance that acts as an inducer. A repressible operon is one whose structural genes will shut off in response to a key metabolic

substance, in this case called a co-repressor. Both function by the reversible binding of a repressor to a DNA sequence that lies between the promoter region and the transcriptional start site. In an inducible operon, the inducer binds a repressor, rendering it incapable of binding to the operator; in a repressible operon, the corepressor binds the repressor, enhancing its ability to bind to the operon.

d. DNA methylation in prokaryotes vs. DNA methylation in eukaryotes

The addition of a methyl group to carbon number five of cytosine serves to protect the DNA of prokaryotes from attack by the cell's own restriction enzymes, present to degrade foreign DNA. In eukaryotes, DNA methylation serves to silence transcription of the methylated regions. Methylation of eukaryotic DNA occurs primarily in gene regulatory regions.

e. liver fibronectin vs. fibroblast fibronectin

These two proteins are actually products of the same gene. They differ in the way that the transcript is processed in the nucleus. Fibronectin from fibroblasts has two peptides missing from the liver protein. These regions of the pre-mRNA are retained in fibroblasts but excised from the transcript in liver cells.

Multiple Choice

1. a	2. c	3. b	4. c
5. a	6. d	7. d	8. b
9. c	10. d	11. b	12. a
13. b	14. c	15 d	16. d
17. b	18. a	19. c	20. d

Problems and Essays

1. a.

	With Lactose			Without Lactose		
Genotype	ß-gal	gal permease	transacet	ß-gal	gal permease	transacet
$i^+ P^+ 0^+ z^+ y^+ a^-$	yes	yes	no	no	no	no
$i^+ P^+ 0^+ z^- y^+ a^+$	no	yes	yes	no	no	no
$i^- P^+ 0^+ z^+ y^+ a^+$	yes	yes	yes	yes	yes	yes
$i^+ P^- 0^+ z^+ y^+ a^+$	no	no	no	no	no	no
$i^+ P^+ 0^- z^+ y^+ a^+$	yes	yes	yes	yes	yes	yes

b. The repressor would not be transcribed due to a mutation in the binding site of RNA polymerase. Thus the structural genes would be expressed whether or not lactose was present.

2. a. Band 1. The DNA of the fragment in Band 1 is not sensitive to DNase I, a sign that it is not being actively transcribed in either tissue.

b. Gel B. The globin gene (and its regulatory regions) are present in Band 2, and sensitive to increasing concentrations of DNase I only in Gel B. This gene is actively transcribed in red blood cells of birds but not in brain tissue.

3. Translation provides a mechanism for amplification of a gene product, because a single transcript can be translated many times. RNA that is not translated, such as rRNA, misses out on that amplification step. In cells with a high demand for ribosomes, such as amphibian oocytes, the DNA that codes for all but one of the rRNA subunits is amplified by selective replication. Thousands of transcriptionally active copies of the genes are produced. The single rRNA subunit that is not selectively amplified, that of the 5S rRNA, is present in the genome in 20,000 copies.

4.

       <u>HLH motif</u>   The transcription factor occurs as a heterodimer with two alpha-helical segments separated by a loop. The alpha helices are perpendicular to one another.

       <u>HLH motif</u>   MyoD

       <u>zinc finger motif</u>   This transcription factor has several zinc ions affiliated with it.

       <u>HMG-box motif</u>   The transcription factor bends DNA resulting in a spatial arrangement that enhances its access to other regulatory factors.

       <u>zinc finger motif</u>   TFIIIA

       <u>HMG-box motif</u>   UBF

5.

       <u>active</u>   Core histones are heavily acetylated.

       <u>inactive</u>   Chromatin is tightly affiliated with histones.

       <u>active</u>   Chromatin is hypersensitive to DNase I.

       <u>inactive</u>   Chromatin is methylated.

       <u>active</u>   Chromatin is affiliated with histones, but the histone octamers are disrupted.

       <u>inactive</u>   Chromatin is in the form of constitutive heterochromatin.

Chapter Thirteen

Short Answer

1. Significance
   a. Meselson and Stahl experiment: *established that DNA replication in prokaryotes is semiconservative.*
   b. temperature-sensitive mutants: *have enabled researchers to study nearly every type of physiological process, including those whose absence would be lethal (such as DNA synthesis).*
   c. the exonuclease activity of DNA polymerase: *5' to 3' exonuclease activity enables the polymerase to remove RNA primers, especially in the lagging strand. 3' to 5' exonuclease activity removes mismatched bases as new DNA is being formed.*
   d. RNA primers: *provide a starting point for DNA polymerases that cannot initiate replication, but elongate only existing nucleotide strands. The initiation of replication occurs when mismatched bases are most likely to be incorporated. Thus RNA primers may also minimize errors, by initiating the process with a temporary strand that is later removed.*
   e. single-stranded binding proteins: *stabilize the single-stranded DNA molecule so that it can serve as a template for replication.*
   f. oriC site: *the one and only starting point for replication in E. coli.*
   g. autonomous replicating sequences (ARSs): *promote replication of the DNA in which they are contained– either the yeast chromosome where they occur naturally, or any DNA in which they are experimentally inserted.*
   h. DNA repair enzymes: *prevent DNA damage, occuring as a result of many natural phenomena, from accumulating and causing mutation and cell death.*

i. intergenic complementation technique: *is used to study mutations that occur in complex, multigenic pathways. By fusing two cells with different mutations in the same pathway, a functional pathway can be reconstructed. Analysis using this technique can establish the number of genes involved in a pathway.*

2. Compare and Contrast:
   a. RNA primers of leading strands vs. RNA primers of lagging strands
      On both the leading and lagging strands, replication depends upon elongation of RNA primers. Due to semidiscontinuous replication, there are many more primers synthesized on the lagging strand than on the leading strand. Also, on the leading strand, RNA synthesis is initiated at the origin of replication by RNA polymerase. On the lagging strand, RNA synthesis begins at the replication fork and is carried out by a different enzyme, called primase.
   b. oriC vs. ter
      Both of these are regions on the bacterial chromosome. OriC is the site of initiation of replication; ter is the site where the two replication forks meet and replication ceases.
   c. base substitution errors in transcription vs. base substitutions in replication
      If a base substitution occurs during transcription, the RNA that results contains an error, and that error may affect all of the polypeptides that are translated from that transcript. But because transcripts are generally short-lived, the error may have little overall, long-term effect on the cell. Base substitutions that occur during replication, on the other hand, become part of the cell's genome. The error is passed down to all cells that arise from that cell and is apparent in all the RNA that is transcribed from that DNA. If the error occurs in a germ cell of a eukaryote, it may even be passed on to the next generation.
   d. replication origins in prokaryotes vs. replication origins in eukaryotes
      Prokaryotes generally have one specific site where replication begins, oriC. Most eukaryotes have many sites, called autonomous replication sequences. In mammalian cells, the origin sites for replication may depend more on features of chromatin organization than on specific DNA sequences.
   e. autonomous replicating sequence vs. origin recognition complex
      These are the two essential components of replication in yeast. The autonomous replicating sequence is the region of DNA that is recognized by a complex of initiation proteins. That protein complex is called the origin recognition complex. Thus it is the binding of the latter to the former that initiates replication.

Multiple Choice

1. a	2. c	3. d	4. c
5. a	6. c	7. d	8. b
9. b	10. a	11. d	12. c
13. b	14. a	15. d	16. a
17. c	18. a	19. b	20. a

Problems and Essays

1.

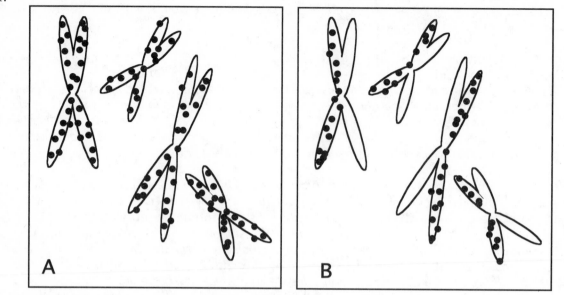

c. *After the first round of DNA replication, each sister chromatid is a combination of one strand from the unlabeled parental DNA duplex, and one newly synthesized strand made with radioactive thymidine. Thus both chromatids will have one labeled strand of DNA. After the second round of replication, only one of the sister chromatids will have a labeled strand of DNA. The second sister chromatid, comprising the unlabeled parental strand from diagram (a) and one newly synthesized strand (also unlabeled), will have no labeled thymidine. Although it is not illustrated here, when experiments like these are actually done, some label appears in the "unlabeled" chromatid during the second cell division. This is because some exchange of genetic material between chromatids occurs during metaphase of mitosis. The corresponding region of the "labeled" chromatid in these cases is free of label.*

2.

3. *Grow your normal E. coli at 37 °, then lyse the cells. Using biochemical techniques, isolate lysate fractions that represent different cellular proteins. You could, for example, isolate fractions of the lysate containing proteins of different molecular weight. Take each fraction of normal cells, and add it to an aliquot of a similar lysate from temperature sensitive cells grown up at 20 °C. Add radioactive precursors to your test vials and test for DNA replication at 37 °C. When you find a fraction of normal cells that restores replication at 37 ° in temperature sensitive cells, perform subfractionations of it until you know which component restores DNA replication.*

316 Appendix

4.

a. Before restriction enzyme digestion:

b. time = 0 minutes:

c. time = 5 minutes:

d. time = 10 minutes:

e. time = 20 minutes:

f. time = 40 minutes:

5.

Natural Bases	Deaminated Bases	Natural Bases	Deaminated Bases

cytosine

uracil

guanine

xanthine

thymine

cannot be deaminated

adenine

hypoxanthine

Uracil is the base that occurs in RNA in place of thymine.

6. *If cytosine becomes deaminated as a result of environmental damage, the base that forms is uracil. Because uracil is not a normal component of DNA, it is recognized and removed by a DNA glycosylase. If, in fact, uracil were the normal base that paired with adenine, as is the case in RNA,*

then DNA glycosylases would not be able to distinguish between uracil that was part of the DNA sequence, and uracil that resulted from deamination of cytosine. Uracil would base pair with adenine, but if that uracil was originally a cytosine, the proper partner would be guanine.

Chapter Fourteen

Short Answer

1. Significance
   a. MPF (maturation promoting factor): *triggers the transition of a cell from the G_2 phase to the M phase, but only when it is activated by cyclin. MPF acts as a cyclin-dependent protein kinase.*
   b. cell cycle checkpoints: *provide points at which the progress through the cell cycle can be halted if conditions are not appropriate for cell division or DNA replication.*
   c. p53 protein: *this transcription factor helps to prevent a cell from becoming malignant by controlling the synthesis of a cyclin-Cdk inhibitor that is required for a cell to enter the S phase of the cycle.*
   d. formation of the mitotic chromosome: *packages the genetic material in a highly condensed form such that its distribution to daughter cells during mitosis is efficient.*
   e. phosphorylation of histones: *plays a role in chromosome packaging prior to cell division by decreasing the affinity of histones for DNA.*
   f. mitotic spindle: *organizes the dividing cell by establishing the poles (at the positions of the centrosomes) and establishing the positions of the chromosomes both before and after separation.*
   g. fragmentation of the endoplasmic reticulum: *facilitates the partitioning of the endoplasmic reticulum into daughter cells during cell division.*
   h. polar microtubules: *form a structural basket that maintains the integrity of the mitotic spindle during cell division.*
   i. slow rate of chromosome movement during anaphase: *ensures that chromosomes segregate accurately, and prevents them from becoming tangled.*
   j. chiasmata: *represent positions along the chromosomes where recombination has likely occurred.*

2. Compare and Contrast:
   a. G_0 period vs. G_1 period of the cell cycle
      These periods of the cell cycle occur before chromosomes have replicated. Cells that are highly specialized and have ceased dividing, such as nerve and muscle cells, are said to be locked in G_0 to distinguish them from cells in the G_1 period that have unreplicated chromosomes but will soon enter the S phase.
   b. centromere vs. kinetochore
      Both of these structures occur at the point of attachment of sister chromatids in a replicated, condensed chromosome. The centromere is part of the chromosome, and forms the site of primary constriction. DNA in this region is characterized by highly repeated sequences that serve as the binding sites for specific proteins. The kinetochore is a platelike structure that occurs at the outer surface of the centromere, and serves as the site of attachment of spindle microtubules. Kinetochores are comprised of proteins and possibly DNA, but the exact composition has not yet been firmly established.
   c. prometaphase vs. metaphase
      These are early stages of mitosis. Prometaphase precedes metaphase, and is characterized by both the attachment of the chromosomes to the spindle fiber microtubules, and the movement of chromosomes toward the equator of the spindle. During metaphase, the chromosomes become fully aligned at the metaphase plate. Metaphase serves as a "checkpoint" for mitosis. The cell will remain at this stage until the chromosomes are properly aligned.
   d. anaphase A vs. anaphase B
      Anaphase is characterized by two separate but simultaneous events: anaphase A and anaphase B. Anaphase A refers to the movement of chromosomes toward the poles, and anaphase B refers to the movement of the spindle poles away from each other.

e. phragmoplast vs. contractile ring

These are both structures that separate the cytoplasm of dividing cells. The phragmoplast in plant cells is a region where microtubules accumulate and eventually form the cell plate. Animal cells divide their cytoplasm by cinching in the cell membrane via actin-containing contractile filaments that form the contractile ring.

Multiple Choice

1.	c	2.	b	3.	a	4.	b
5.	d	6.	a	7.	c	8.	a
9.	c	10.	d	11.	c	12.	a
13.	b	14.	b	15.	c	16.	a
17.	b	18.	a	19.	c	20.	c

Problems and Essays

1.

Experiment 1   Experiment 2

In experiment 1, the active MPF in the cytoplasm of the S cell would induce DNA replication in the G_1 nucleus, and thymidine would be incorporated into the G_1 chromosomes. In experiment 2, the G_2 nuclei could not be induced to undergo a second round of replication, hence no radioactive thymidine would occur in the G_2 chromosomes.

2. First, isolate the DNA from the human cells, and digest it into fragments, using restriction endonucleases. Separate the fragments, using gel electrophoresis, and then introduce each fragment into a different culture of temperature-sensitive yeast mutants. Try growing your yeast mutant cultures with the human DNA, at both the permissive temperature and the nonpermissive temperature. The yeast cells that are able to grow at the nonpermissive temperature have been "rescued" by the human DNA and must contain a functional gene for the mutated cell cycle protein. Now isolate the DNA from your rescued yeast, and separate the human DNA from those cells. That DNA has an essential cell cycle gene– one that encodes a cell cycle control protein.

3.

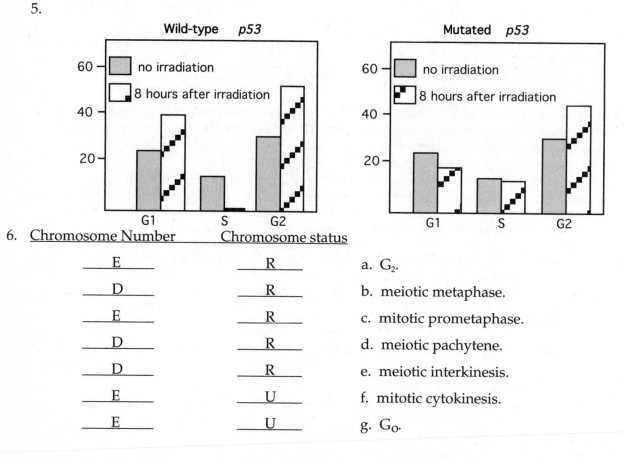

cyclin concentration

MITOSIS

MPF activity with RNase

MPF activity

MITOSIS

Time ➔

4. a. *Mutant 1 is deficient in the protein coded by the gene cdc 25, hence it is the cdc 25⁻ mutant. The cdc25 activates the cdc2 by removing inhibitory phosphates. Cells grow abnormally long if the cdc25 gene is not producing a functional phosphatase. Mutant 2, on the other hand, is deficient in the protein coded by the wee1 gene. Wee1 codes for a kinase that adds the inhibitory phosphates to cdc2. In the absence of wee1, cells divide prematurely.*

   b. *In both mutants, the duration of the G_2 phase is abnormal. In mutant 1 it is lengthened, and in mutant 2 it is shortened.*

5.

Chromosome Number	Chromosome status	
E	R	a. G_2.
D	R	b. meiotic metaphase.
E	R	c. mitotic prometaphase.
D	R	d. meiotic pachytene.
D	R	e. meiotic interkinesis.
E	U	f. mitotic cytokinesis.
E	U	g. G_0.

6.

7. a.

enlarge this
area

b.

Chapter Fifteen

Short Answer

1.   Significance
 a. the many steps of a signal transduction cascade: *provide many potential sites for signal amplification, as well as potential sites for regulation of the signal.*
 b. cAMP: *acts as a second messenger, activating protein kinase A in response to ligand binding to a membrane-bound receptor on the external cell surface.*
 c. phosphatases: *remove phosphates from proteins. Depending on the substrate protein, dephosphorylation may activate or inactivate the substrate, and terminate the intracellular signal.*
 d. calmodulin: *binds up to four calcium ions. In the bound form, calmodulin can activate protein kinase C, phosphodiesterase, or calcium transport.*
 e. SH2 domain: *endows a protein with high affinity for phosphotyrosine motifs on IRSs. Proteins with these domains bind to IRSs, and their activity is modified in a way that alters cell differentiation and growth.*
 f. receptor tyrosine kinases: *phosphorylate tyrosine residues on substrate proteins (IRSs) upon binding their specific ligands at the cell surface. Phosphorylation results in changes in cell growth and differentiation.*
 g. Ras protein: *the protein that acts as the focal point for several important signal cascades involved in cell differentiation and growth. Mutations in the ras gene are implicated in 30 to 40 percent of human cancers.*
 h. MAP kinase cascade: *important (and widespread) signal transduction pathway involved in regulation of transcription, replication, and translation.*
 i. crosstalk: *enables different signal transduction pathways to function in a coordinated and interdependent fashion.*
 j. apoptosis: *rids the organism of cells that are no longer needed or may be potentially harmful to the individual.*

2.   Compare and Contrast:
   a. kinase vs. phosphatase
      Both types of enzymes involve phosphate groups, and both play important roles in intracellular signal transduction pathways. Kinases transfer phosphate groups to tyrosine, serine, or threonine residues on target proteins, whereas phosphatases remove phosphate groups from those residues on target proteins.
   b. heterotrimeric G proteins vs. monomeric G proteins
      These proteins both function as signal transduction molecules, relaying information from a ligand-bound receptor at the cell surface to an intracellular effector molecule. Both are bound to GTP in the active form, and both have GTPase activity. Upon hydrolysis, both GDP-bound G proteins are switched off. Heterotrimeric G proteins function by dissociation of the α subunit which interacts with an effector. Monomeric G proteins, such as ras, activate effectors. When a ligand binds to a receptor tyrosine kinase and causes autophosphorylation, SH2 proteins are recruited, and, via an sos intermediate, the G protein releases GDP and binds GTP.
   c. phospholipase C vs. protein kinase C
      Both of these enzymes function in signal transduction pathways, but their substrates and actions are quite different. Phospholipase C, activated by the α subunit of a heterotrimeric G protein following, for example, the binding of acetylcholine to its receptor, removes the inositol phosphate moiety from phosphatidylinositol in the membrane, creating IP3 and DAG, both of which function as second messengers. Protein kinase C, activated by DAG and sometimes calcium, transfers a phosphate group to serine or threonine residues on target proteins.
   d. tyrosine kinases vs. serine/threonine kinases
      Each of these kinases transfers phosphate groups to target proteins, but to different sites on the target proteins. Most kinases are specific for either tyrosine residues or serine and/or threonine residues.
   e. signaling pathways in plants vs. signaling pathways in animals
      Basically, members of these two kingdoms use quite similar signal transduction pathways, with a few exceptions. Whereas both groups use changes in intracellular calcium, IP3, and DAG, only animals use cyclic nucleotide monophosphates as second messengers, and only plants use salicylic acid. Histidine kinases appear to be specific to plants, as well.

Multiple Choice

1.	b	2.	d	3.	b	4.	d
5.	b	6.	b	7.	a	8.	c
9.	d	10.	c	11.	d	12.	b
13.	d	14.	c	15.	b	16.	a
17.	c	18.	b	19.	d	20.	c

Problems and Essays

1.

Treatment	Additions	DNA Replication
a. none	PDGF or EGF	*yes*
b. microinjection (intracellular) of constitutively active Ras protein	none	*yes*
c. microinjection of saline ("mock" injection)	PDGF or EGF	*yes*
d. microinjection of saline ("mock" injection)	none	*no*
e. microinjection of anti-Ras antibodies	PDGF or EGF	*no*
f. microinjection of anti-Ras antibodies	none	*no*

2. a. √_____
    b._____
    c._____
    d. √_____
    e. √_____
    f._____

3. *The Ras protein is a small monomeric G protein that is turned on by binding GTP and turned off by hydrolyzing GTP to GDP. If the ras gene is mutated such that the GTPase activity is lost (in this case due to a single amino acid substitution), Ras is never inactivated, and the signal cascade is always on. Transcription, translation, replication, and hence growth and division proceed unregulated. The result is malignancy.*

4. *Because four moles of GppNp bind to 0.8/100 moles of receptors, then 500 GppNp bind per receptor. Thus the extra 500-fold amplification must occur at the level of the receptor-G protein interaction.*

5. *The serotonin receptor is a membrane-bound, heterotrimeric G-protein coupled receptor whose effector is adenylyl cyclase.*

Chapter Sixteen

Short Answer

1. Significance
    a. aneuploidy in transformed cells: *a sign that the genome of cancer cells is not being carefully maintained, and that the growth of cancer cells is not as dependent upon a normal chromosome complement as the growth of normal cells.*
    b. Percival Pott's studies of British chimney sweeps: *the first studies to link cancer to environmental factors– in this case, soot.*
    c. tumor-suppressor genes: *code for proteins that keep the cell cycle in check. Mutations in both copies of these important genes result in uncontrolled cell growth and reproduction.*
    d. proto-oncogenes: *code for proteins involved in many aspects of the cell cycle, including the signal transduction pathways that initiate cellular reproduction. Malfunction of either allele for these genes, as a result of mutation, inappropriate transcriptional control, or infection by viruses, transforms them into oncogenes.*
    e. chromosomal deletion in children with retinoblastoma: *occurs in the region of the chromosome containing a gene, RB, that plays an important role in down-regulating the cell cycle. This is a clear indication that the condition can be inherited.*
    f. the oncogene, sis, of the simian sarcoma virus: *the first oncogene to be identified as coding for a growth factor, namely PDGF.*

2. Compare and Contrast:
    a. benign vs. metastatic tumor
    Both types of tumors result from uncontrolled cell growth, but benign tumors lack the ability to metastasize to distant sites in the body and are therefore contained and treatable by surgery. Metastatic tumors occur when cells break away from the original tumor and invade adjacent or distant sites, growing uncontrollably at these new sites. Metastatic tumors often begin as benign tumors.
    b. tumor virus genes vs. oncogenes
    Both of these stretches of DNA code for proteins that cause the cell to lose control of growth and reproduction. Tumor virus genes enter the genome of the host cell by viral infection, inserting the oncogenic DNA into the host DNA, transforming the host cell to a cancer cell. The term oncogene is a more general term, referring to any gene (whether it be of viral origin, or a mutated or malfunctioning part of the normal cell genome) that transforms a cell from normal to malignant.

c. initiators vs. promoters

These terms both refer to environmental agents that play a part in transforming cells from normal to malignant. The initiator causes the original mutation that results in the creation of an oncogene. The promoter is the agent that causes the cancer to be expressed, either by creating conditions of cell growth that favor the development of other mutations or via other epigenetic mechanisms.

d. tumors in children with familial retinoblastoma vs. those with sporadic retinoblastoma

Both of these cancers result from the malfunction of the product of the RB gene. If the condition is familial, i.e., inherited as a deletion in one member of chromosome 13, the tumors tend to be multiple and in both eyes. The other member of the missing allele on chromosome is invariably mutated. If the condition arises sporadically, it is usually characterized by one tumor in one eye, and while there is no chromosomal deletion, both alleles of the RB gene are found to be mutated.

e. oncogenes that encode growth factors vs. those that encode growth-factor receptors

Any oncogene will result in a cell that has lost normal controls over its growth and reproduction. When a gene coding for a growth factor is overexpressed, uncontrolled growth results from an overabundance of the growth factor in the cell's environment. If the gene coding for the receptor is mutated, the growth-factor signal may be constitutively "on," regardless of whether the growth factor is there to turn it on.

Multiple Choice

1. d	2. a	3. b	4. d
5. b	6. c	7. b	8. b
9. d	10. b	11. a	12. d
13. c	14. a	15. d	16. d
17. b	18. b	19. c	20. d

Problems and Essays

1. a. chemotherapy: *Treating cancer patients with high doses of toxic chemicals is designed to selectively kill cancer cells by exploiting their higher sensitivity to these toxins. For many cancers, doses that are sufficient to eradicate the cancer are too toxic for the survival of normal cells, and either the cancer is not eliminated, or the patient dies.*

b. gene therapy: *Introducing functioning genes into cells that have lost control of growth and reproduction is a possible therapy because in cancer patients, proteins that control these functions are coded by mutated genes. While this approach may someday be the treatment of choice, so far it has been impossible for the new and functioning bits of DNA to reach billions of cancer cells buried deep within tissues.*

c. surgery: *Removing cancer cells from the body surgically would cure the disease if the cancer were caught before it had a chance to metastasize. Unfortunately, that is not always the case, and it is not always possible to determine if a cancer has metastasized.*

d. stimulating the immune system: *This approach, which recruits the body's own immune surveillance system into detecting and destroying cancer cells, was a source of great optimism in the early 1980s, but has not been as successful as was hoped.*

e. anti-angiogenic compounds: *Preventing the growth of new blood vessels to tumors deprives them of the nutrients they need to survive. In addition, tumors do not become resistant to these agents as they do to chemotheraputic agents. This approach holds great promise, but for reasons that are not yet clear, not all patients respond.*

2. $$\frac{6 \times 10^9 \ people \ \times \ 100 \times 10^{12} \ cells/ \ person}{50 \ years} \times \frac{1}{3} = 4 \times 10^{21} \ malignant \ cells \ per \ year$$

3. a. *Isolate the RNA from your strain of normal, nonmutated tumor virus. Using reverse transcriptase and radiolabeled deoxyribonucleotide triphosphates, synthesize complementary DNA (cDNA) in vitro. Isolate the RNA from the mutated tumor virus, and hybridize it to the cDNA from the normal virus. Pass the hybridized nucleic acid over a hydroxyappatite column to separate the DNA-RNA hybrids from the unhybridized DNA. The unhybridized DNA represents part or all of the gene that is responsible for causing murine leukemia.*

b. *The first step is to determine the sequence of the cDNA involved in the cancer, and therefore the amino acid sequence of the gene product. One good strategy for determining function involves comparing this sequence to those of genes of known function, and looking for particular amino acid domains or motifs that occur in an oncogene product whose function can be identified. For example, does the protein contain a zinc finger motif? How about a tyrosine kinase domain? Comparisons of this kind give clues about the function of the protein. The sequences of nucleic acids and amino acids for many different proteins are available on data bases accessible to students and researchers in this field. Ask if your university can access these gene bank data bases.*

4. a and b:

c. *Cells in the normal +EGF and -EGF cultures are arrested in the Go phase of the cell cycle after day four.*

5. *If a mutation occurs in one allele of a tumor-suppressor gene, the product of that gene will not function properly, but the malfunctioning protein, by itself, does not cause malignancy. The cause of cancer is the absence of a properly functioning tumor-suppressor protein. The gene product of the other allele, which is not mutated, can assume the role in cell cycle regulation of that locus. Thus a mutation in a tumor-suppressor gene acts as a recessive. In the case of an oncogene, the presence of the malfunctioning protein is enough to cause cancer. The action of the gene product is to stimulate cell growth and division, not down-regulate it. Thus even in the presence of a properly functioning proto-oncogene product, which would be produced from the allele that did not undergo mutation, the malignant phenotype would be expressed. The gene would act as a dominant.*

6. *A cancer initiator causes the first (and sometimes only) mutation in a sequence of events that leads to cancer. Mutations are permanent features of the genome; they do not fade or heal with time. Tumor promotion usually results from stimulated growth or some other condition whereby the likelihood of a second mutation (and full-blown cancer) is increased. The effects are cumulative over time, and further, the passage of time cannot mitigate the permanent nature of initiation.*

7. a. *Cell line A is transformed. The p53 gene is probably involved in the transformation.*
   b. *Cell line B is normal.*
   c. *Cell line C is transformed. Based on the description, it is not possible to identify the oncogene, but it is probably not the p53 gene because apoptosis relies on a functioning p53 protein.*
   d. *Cell line D is transformed, and p53 is undoubtedly the oncogene involved.*

Chapter Seventeen

Short Answer

1. Significance
   a. memory cells: *ensure a rapid secondary immune response. These activated B cells remain in the lymphoid tissue, some for the life of the individual, where they can be stimulated to proliferate should the antigen that originally activated them be encountered again.*
   b. clonal selection: *explains the origin of the specific acquired immune response via the activation of only lymphocytes making certain antibodies or TCRs from a pool of lymphocytes with many different specificities.*

c. professional APCs: *primarily macrophages and dendritic cells that engulf antigens, process them by intracellular digestion, and exhibit fragments of them on their surface. Helper T lymphocytes recognize the antigen fragments and activate either CTL cells or B cells to eliminate the antigen.*

d. multiple myeloma as it pertains to research in immunology: *antibodies in the blood of any one of these patients suffering from the disease are largely monoclonal, but different patients produce different monoclones. Antibodies from these patients provided important insights into the structure of antibodies, including the presence of constant and variable domains, and aspects of gene rearrangements in antibody production.*

e. a single antigen may have several different epitopes: *therefore, any one antigen will have more than one clone of antibodies directed against it; hence the immunological response is more effective because it is polyclonal.*

f. somatic hypermutation: *acts as a source of variability in the genes encoding immunoglobulin proteins. This process can improve the affinity of the antibodies produced by a cell lineage to its antigen.*

g. Ii protein in the lumen of the rough endoplasmic reticulum: *shields the peptide binding site on MHC II preventing it from binding to an endogenous protein until it enters a lysosome or endosome, where MHC II will encounter the fragments of extracellular antigens, to which it binds.*

h. all cells express MHC I: *a mechanism for any cell that could potentially become the victim of an intracellular pathogen to display fragments of peptides on its surface. If the cell is a victim, the fragments will identify it as such and mark the cell for destruction by CTL.*

i. costimulatory signal: *protects the body from autoimmune attack by T_H cells. Only professional APCs that have ingested and processed antigen can provide the costimulatory signal that is required to activate T_H cells, thus ensuring that other cells displaying proteins capable of binding to TCRs will not activate T_H.*

j. JAK-STAT pathway: *stimulates transcription of interferon-stimulated genes in cells that recognize and bind cytokines.*

2. Compare and Contrast:

a. innate vs. acquired immunity
Both of these terms describe responses of the body to invasion by a pathogen, but they differ in important ways. The innate response is rapid and nonspecific, providing a first line of defense. Acquired immune responses are highly specific defenses directed against antigen molecules on the pathogen. They require a lag period after the first encounter, but can be mounted quite rapidly if the same pathogen is encounted at a later time.

b. humoral vs. cell mediated immunity
Both types of acquired immunity involve lymphocytes that produce proteins capable of recognizing and binding to antigenic epitopes on foreign matter. Humoral immunity is mediated by antibodies produced by B lymphocytes and is directed against extracellular antigens. Cell-mediated immunity involves cell surface proteins on T lymphocytes that recognize and bind antigenic fragments displayed on the surfaces of infected cells, or on the surfaces of professional antigen presenting cells, such as phagocytes or dendritic cells. Cell-mediated immunity can rid the body of intracellular pathogens.

c. cytotoxic T cells vs. helper T cells
These two major classes of T lymphocytes are both involved in cell-mediated immunity; helper T cells are also involved in humoral immunity. Differences in their cell surface proteins reflect the different roles they play. Both types of T cells have T cell receptors that recognize MHC-antigen complexes, but cytotoxic T cells recognize MHC I complexes on the surfaces of infected cells and release cytotoxins that kill infected cells displaying antigen fragments. Helper T cells recognize MHC II complexes on the surfaces of APCs and are activated by binding with these complexes. Activated helper T cells release IL molecules that stimulate B cells to mount a humoral response, and also stimulate cytotoxic T cells to kill cells displaying MHC I-antigen fragment complexes.

d. antibodies vs. T cell receptors

These two classes of proteins are capable of specific binding to antigens. Antibodies are the immunoglobulin proteins produced by B cells and B cell progeny (plasma cells) that are released into the blood where they affect humoral immunity against extracellular pathogens. T cell receptors occur on the surface of T cells where they recognize and bind to antigen fragments displayed by the MHC complexes on the surfaces of infected cells and professional antigen-presenting cells. T cell receptors are important in antigen recognition during cell-mediated immunity via cytotoxic T cells. They play an indirect role in humoral immunity on helper T cells when they are activated by APCs, which, in turn, activate B cells.

e. MHC I vs. MHC II

These proteins of the histocompatibility antigen group are highly polymorphic proteins that display fragments of processed antigens on the surfaces of cells. MHC I is produced by nearly all cells of the body, and can display both self and non-self proteins. Cells with MHC I displaying non-self proteins are targeted for destruction by cytotoxic T cells. MHC II is produced primarily by professional antigen presenting cells, such as macrophages and dendritic cells, where it displays fragments of exogenous proteins taken up by the APC. MHC II, in conjunction with other proteins, binds to helper T cells and activates the specific immune response.

Multiple Choice

1. b	2. a	3. d	4. d
5. a	6. a	7. d	8. c
9. a	10. c	11. c	12. a
13. c	14. b	15. a	16. d
17. b	18. c	19. d	20. a

Problems and Essays

1. *Solid line A represents the antibody titers when passive immunization is used to fight an antigen that has already entered the body. Because passive immunization involves injecting exogenous antibodies, there is no lag time between injection and an increase in antibody titers, but neither is there any lasting immunity. The body is not producing the appropriate antibody, and when the injected antibody proteins turn over, immunity is lost. Dashed line B represents the antibody titers following vaccination. The initial response is slower than passive immunization, but subsequent encounters with the antigen stimulate a rapid proliferation of antibody-producing cells without additional injections. Vaccination stimulates endogenous antibody production and memory cells, conferring lasting immunity.*

2.
   S a.
   N b.
   S c.
   I d.
   S e.

3.

Antibody	B cells	Helper T cells	CTL	Macrophages
anti-MHC II	+			+
anti-CD4		+		
anti-CD 8			+	
anti-TCR		+	+	
anti-MHC I				
anti-IgG	+/-*			
anti- CD28		+		
anti-B7				+

*B cells will have antibodies on their surface, but they may or may not be of the G class of immunoglobulin.

4.   To test for genetic rearrangement during class switching you would need DNA from B cells producing IgM early after activation, and also from B cells producing IgG with the same antigen specificity later after activation. These two types of cells could be separated from a monoclonal culture by "panning for cells" as described in question 3. The two DNA preparations are digested with restriction endonucleases, and DNA fragments that contain the C_H genes for IgG and the VDJ genes are hybridized with radioactive probes that bind to these sequences, i.e., a Southern blot. If class switching involves genetic rearrangement, the fragments that hybridize to these probes will look different after class switching. Depending on the restriction enzymes used, they may be on separate fragments before class switching and on the same fragment after class switching.

5.   Group 1 Treatment: _These mice were exposed to antigen Z at least several weeks prior to the first measurement of serum antibody titers._

Explain: *There is a low but constant level of antibody to antigen Z in the serum of these mice indicating that at some point they were exposed to the antigen. The absence of a peak, however, indicates that antibody production has not been stimulated by exposure to the antigen in the most recent several weeks.*

Group 2 Treatment: _This is the experimental control; they have never been exposed to the antigen._

Explain: *The absence of detectable antibodies for antigen Z indicates that these mice have not been exposed to the antigen.*

Group 3 Treatment: _First exposure to the antigen at approximately week 0, and second exposure sometime between weeks 3 and 5._

Explain: *The initial peak in antibody titers during week 2 is typical of a primary immune response. The much larger second peak indicates a second exposure that elicited a secondary response characterized by activation of memory cells.*

6.   ___K___ a.

Explain: *The CTL recognized the antigen as foreign, and the MHC surface proteins on the fibroblast as compatible (because the fibroblasts were from genetically identical mice); hence the CTL killed the fibroblasts.*

<u>NK</u>   b.

Explain: *The fibroblasts were not displaying antigen and hence CTLs were not activated to kill. This is one kind of control protocol for such a series of experiments.*

<u>NK</u>   c.

Explain: *Whereas the antigen displayed on the fibroblast is "familiar" to the CTL, the MHC protein probably is not because the CTL and the fibroblasts are not genetically identical. There is one exception: If both the CTL donor strain and the fibroblast donor strain were identical at the loci encoding the MHC proteins, then CTL could kill the infected fibroblasts. The high level of polymorphism at these loci, however, makes this unlikely.*

<u>NK</u>   d.

Explain: *Unless the mouse from which CTL were taken had encountered virus B previously, the CTL population from the spleen would have very few cells capable of recognizing the virus. A few fibroblasts would be killed, but not many.*

<u>NK</u>   e.

Explain: *The CTL would recognize the antigenic virus A, but the MHC proteins would not be compatible due to the mutation at the MHC I locus.*

Chapter Eighteen

<u>Short Answer</u>

1.   Significance
   a. microscopic resolution: *determines the fine detail that can be seen by a given microscope under given conditions.*
   b. interference microscopy: *makes transparent objects more visible. This technique can be used to examine unstained living cells that would be impossible to see in bright-field microscopes.*
   c. electron microscopy: *provides vastly greater resolution than any form of light microscopy by using a beam of electrons instead of a beam of light. This technique was instrumental in determining the fine structure of cells*
   d. autoradiography: *when it is applied to whole cells, identifies the intracellular location of radiolabeled components. When it is applied to gels or agar plates, this technique identifies the bands or colonies that are radioactive.*
   e. cell culture: *removes cells from the complex factors that influence them as part of multicelluar organisms, and allows them to be studied under simplified conditions. This technique has proven to be one of the most successful and important advances in cell biology.*
   f. two-dimensional gel electrophoresis: *separates complex mixtures of proteins on the basis of both molecular weight and isoelectric point. This high-resolution technique can separate thousands of proteins from the same mixture.*
   g. nucleic acid hybridization: *is used to isolate or identify specific DNA fragments by base-pairings between complementary DNA or RNA probes and the DNA of interest. This technique can also be used to measure the similarity between two DNA sequences.*
   h. yeast artificial chromosomes: *a technique for cloning and storing DNA fragments that are larger (100 to 1000 kb) than any that can be cloned using plasmids. This has been instrumental in the success of the Human Genome Project thus far.*
   i. animal models for human disease: *provide a laboratory system for studying aspects of human disease that would be impossible or unethical to study in humans.*
   j. polymerase chain reaction: *rapidly amplifies minute quantities of DNA. This technique is more time- and cost-effective than cloning, and it has virtually revolutionized DNA technology.*

2. Compare and Contrast:
   a. magnification vs. resolution
      These terms both apply to the ability of the microscope to make small specimens visible. Magnification refers to the ratio of the size of the image formed by the microscope to the actual size of the specimen. Resolution is the distance between two points that can be seen as distinct entities. Magnification aids in resolution, but does not solely determine it. Both are necessary to see subcellular structure.
   b. transmission electron microscopy vs. scanning electron microscopy
      For magnification and resolution of small structures, these two techniques depend upon a beam of electrons altered by a specimen. TEMs form images from electrons that are transmitted through a specimen and then focused on a screen or photographic plate. SEMs depend on the image produced when electrons bounce off the surface of a specimen and then onto a screen or photographic plate. TEMs give high resolution of small intracellular structures seen through thinly sectioned specimens. SEMs reveal surface features of unsectioned specimens.
   c. differential centrifugation vs. density-gradient centrifugation
      Both techniques separate and resolve particles in suspension by subjecting cellular homogenates to centrifugal force. Differential centrifugation is usually used to separate organelles and large subcellular particles, all of which are more dense than the medium. Density-gradient centrifugation can also be used to separate organelles and large particles, but is often applied to small particles or macromolecules. In density gradient centrifugation, the medium forms a gradient, and the particles have densities that are less than the medium at the bottom of the tube. Particles sediment from the top of the gradient to the region of equal density within the medium.
   d. cloning eukaryotic DNA in bacterial plasmids vs. in phage genomes
      Both of these are techniques for amplifying and storing DNA fragments from eukaryotic sources. Bacterial plasmids are commonly used vectors appropriate for cloning tens to hundreds of DNA fragments. For larger DNA samples, viral vectors are better. Bacterial "lawns" on petri plates can accommodate hundreds to thousands of viral plaques, and each one can be screened for the gene of interest.
   e. cloning genomic DNA vs. cloning cDNA
      Both types of clones can be used to amplify and isolate genes of interest. Genomic DNA includes regulatory sequences and intervening sequences whereas clones using cDNA include coding sequences only. Purified cDNA is good for sequencing a gene, and determining the amino acid sequence of a protein. Clones of genomic DNA can provide information about the evolution of DNA, members of gene families, and the ways in which genes are regulated.

Multiple Choice

1. c	2. b	3. a	4. b
5. a	6. d	7. d	8. b
9. a	10. d	11. c	12. c
13. b	14. c	15. b	16. c
17. d	18. c	19. a	20. d

Problems and Essays

1. a.

$$D = \frac{0.61 \times \lambda}{n \sin \theta}$$

$$D = \frac{0.61 \times 527 \text{ nm}}{1.0 \sin 70°}$$

$$D = 342 \text{ nm}$$

b.

$$D = \frac{0.61 \times \lambda}{n \sin \theta}$$

$$D = \frac{0.61 \times 450 \text{ nm}}{1.0 \sin 70°}$$

$$D = 292 \text{ nm}$$

c.

$$D = \frac{0.61 \times \lambda}{n \sin \theta}$$

$$D = \frac{0.61 \times 450 \text{ nm}}{1.5 \sin 70°}$$

$$D = 195 \text{ nm}$$

The resolution in part ©, 195 nm, is about the best you can expect from a light microscope.

2. a. *Whole homogenate, Supernatant 1, and Pellet 2*
   b. *Pellet 2*
   c. *Whole homogenate*
   d. *Supernatant 3*

3. *Amino acids are linked together starting with the amino end and ending with the carboxyl end. In Dintzis' experiment, the first fragments to incorporate radiolabel were those at the carboxyl terminal. Those peptides were nearly finished when the labeled amino acids were added, and therefore only this last segment of the peptide incorporated labeled amino acids. The last fragments to incorporate the label were those at the amino terminal, indicating that their synthesis had begun after the labeled precursors were added.*

4. a.
$$400,000 = \frac{(\omega^2 r) \ cm/sec^2}{980 \ cm/sec^2}$$
$$(\omega^2 r) = 3.92 \times 10^8$$

   b.
$$v = s(\omega^2 r) \ where \ s = 18 \times 10^{-13} \ sec$$
$$v = (18 \times 10^{-13} \ sec) \times (3.92 \times 10^8)$$
$$v = 7.06 \times 10^{-4} \ cm/sec$$

   c.
$$\frac{7 \ cm}{(7.06 \times 10^{-4})(3600 sec/hr)}$$
$$= 2.76 \ hr$$

5. *Gene to Protein:*
   a. *Isolate the gene of interest from genomic DNA.*
   b. *Use a clone of the genomic DNA to probe a population of mRNAs.*
   c. *Make a cDNA using the mRNA for the gene of interest.*
   d. *Sequence the cDNA, and use its sequence to deduce the amino acid sequence of the protein.*
   e. *Compare the amino acid sequence with that of other proteins whose functions are known.*
   f. *Express the gene using a plasmid vector in E. coli or some other expression vehicle.*

Protein to Gene:
   a. *Isolate the protein by using either its molecular weight, its isoelectric point, or its function.*
   b. *Sequence the amino acids of at least part of the protein.*
   c. *Deduce the nucleotide sequence of at least part of the gene that codes for the protein.*
   d. *Synthesize a radiolabeled oligonucleotide to use as a probe to search the genomic DNA for the gene.*
   e. *Isolate the entire gene and determine its sequence (and that of regulatory elements).*

6.

7. a. *polymerase chain reaction followed by restriction enzyme digestion and gel electrophoresis to search for restriction fragment length polymorphisms*
   b. *radiolabeled carbohydrate tracers followed by autoradiography*

c. *nucleic acid hybridization between a radiolabeled probe for growth hormone, followed by gene isolation using restriction enzyme digestion, recombinant DNA, and cloning, and then protein purification from cultures of cells genetically engineered to express the hormone*

d. *site-directed mutagenesis followed by enzyme kinetics*

NOTES

NOTES

NOTES

NOTES

NOTES

NOTES

NOTES

NOTES

NOTES

NOTES

NOTES

NOTES

NOTES